The "Connaught" Two-Stroke Motor Bicycle with 3-speed Gear and Free Engine. As Ridden by the Author.

THE
MotorCyclist's Handbook

A Practical Manual on the Construction and Management of Motor Cycles

(WITH A CHAPTER ON CYCLE-CARS),

By "PHŒNIX"
(Chas. S. Lake).
Author of "How to Drive a Motor Cycle," etc.

FULLY ILLUSTRATED.

THIRD EDITION.
REVISED AND ENLARGED.

©2007-2011 Periscope Film LLC
All Rights Reserved
ISBN #978-1-935700-55-5
www.PeriscopeFilm.com

THE MOTORCYCLIST'S HANDBOOK.

CONTENTS.

PREFACE

TO THE FIRST AND SECOND EDITIONS.

IT has been said, with truth, that an inherent love of things mechanical finds a more or less definite place in the character of every Englishman, even though in many it may be but a latent sense, unassociated either with their business pursuits or pleasures, and there can be little doubt that the motor cycle—which has increased in popularity by leaps and bounds during the past few years—provides a direct and powerful means of fostering that spirit of mechanical aptitude which so many possess without knowing it.

To own a motor cycle is, in effect, to own a private loco-motive, capable of transporting its rider up hill and down dale for long distances, with a rapidity impossible with any other class of vehicle of the same size and weight. The delightful sensation of free-wheeling down a long and easy grade on a pedal cycle—an experience which most of us frequently enjoy—is reproduced, with a motor cycle, on the level, up steep hills, and, indeed, in every stage of progress, calling for no effort on the rider's part, and requiring no particular skill in order that good results may be obtained.

The modern motor cycle is both easy and comfortable to ride and drive ; its construction is relatively simple, and, indeed, every detail connected with its control and mainten-ance may be completely mastered by anyone of average intelligence in a very short time. In engine power it ranges from the little $1\frac{1}{2}$ h.-p. lightweight to the heavy and powerful 7–9 h.-p. twin-cylinder machine, equipped with two- or even three-speed gear mechanism, and suitable for conveying two persons, by means of a side or fore car attachment, from one end of the country to the other, with but very little, if any, difficulty, whatever may be the nature of the hills or the general character of the roads traversed.

Between these two extremes there are to be found many different types of motor cycles, varying in engine-power, price, and constructional details, and, in fact, anyone who, having once made up his mind to become a motor cyclist, and who is prepared to expend a reasonable sum in acquiring a machine suitable to his tastes, need not, now that the motor bicycle has attained its present state of reliability and general efficiency, hesitate for a moment about taking the step which will bring him into personal acquaintance with one of the most delightful pastimes ever introduced.

In the author's experience, however, as contributor of the "Motor Cycle Notes" appearing weekly in the pages of *The Model Engineer and Electrician,* there *does* exist—as is shown by the many inquiries received—a considerable doubt on the part of many as to whether the use and maintenance of a motor cycle does not impose a heavy expenditure in time and money on its owner, and also whether the mechanical ability required in the upkeep of such a machine is not rather in advance of that possessed by the average person. The author's reply to all such queries is a decided negative : For although it would, of course, be idle to assert that motor cycling is one of the cheapest pastimes, the expenses attending it may, if care be exercised, be kept within quite moderate limits ; and while in the early stages of his experience every motor cyclist meets with difficulties which for the moment may appear formidable, the surmounting of these difficulties often provides one of the most interesting and attractive features of the pastime, and fits the rider for tackling any kind of trouble which may subsequently fall to his lot. It is with a view to clearing away such misconceptions as appear to prevail among those who have not yet joined the ranks of the motor cycling community, by placing before them, in as simple and straightforward a manner as possible, the anatomy of the motor bicycle and a description of the functions and formation of its various parts, as well as the uses and capabilities of the machine as a whole, that the preparation of the present work was undertaken—a task, be it known, that owes its conception in some measure to the expressed wishes of many readers of *The Model Engineer* Notes.

"PHŒNIX."

PREFACE TO THE THIRD EDITION.

WHEN, during the year 1911, the author entered upon the task of writing a book on the subject of motor cycles, it was without any anticipation of having to revise the work for a second edition within a few months of the time at which the first edition, numbering some 5,600 copies, made its appearance. Still less was it considered probable that within eighteen months or so from the sale of the first copy, nearly 12,000 would have been disposed of, and that the publishers would, by then, be requiring a revision of the work, on a large scale, for the third edition.

Developments associated with both motor cycles and motor cycling have, however, been remarkably rapid during the past two seasons. Manufacturers have improved their designs, and the popularity of motor cycling has increased by leaps and bounds amongst the general public. These circumstances, coupled with the ready demand experienced for it, have made it necessary to adopt a similar attitude towards " The Motor Cyclist's Handbook," and, by introducing new matter, descriptive of the latest developments in the various constructional features of the motor cycle, both separately and collectively (with new illustrations to aid in their representation), bring the work into line with the most modern standards and requirements, and so extend its scope and the sphere of its usefulness. Certain chapters have been enlarged, and a new chapter on cycle cars added.

The pages of the book, in its present edition, will be found to contain information bearing upon features of motor cycle construction and equipment which, when the previous editions were issued, had either not been introduced, or were in the earliest experimental stages, and while, for obvious reasons, it would neither be possible nor politic to treat *only* of the very latest practice in every direction, sufficient has, it is hoped, been done to make the book equally useful to those who require information bearing chiefly upon the most recent developments and those who wish for a more general insight into the subject as a whole.

The author takes this opportunity of thanking the many readers of the first and second editions who have been good enough to write expressing their appreciation of the work

" PHŒNIX."

ERRATUM.

On page 22, last line, *for* Fig. 192, page 245, *read* Fig. 182, page 234.

The Motor Cyclist's Handbook.

By " Phœnix."

Contributor of the " Motor Cycle Notes " appearing weekly in *The Model Engineer and Electrician.*

CHAPTER I.

The Engine : Its Working Principle ; the Four-stroke Cycle ; Valves, and the Means of Operating Them.

As in every other mechanically driven vehicle, the motor cycle depends for its means of propulsion upon a power unit, which in this case consists of an internal combustion engine consuming petrol gas as its fuel, and working, in the majority of instances, upon the four-stroke, although, in a few recent designs, upon the two-stroke (or double-impulse) principle. The prospective motor cyclist, if he be wise, will first make himself acquainted with the construction and system of working of the engine before setting out to acquire possession of a motor cycle, so that whatever difficulties he may encounter during his novitiate may be the more readily and independently overcome. Subsequent personal experience and acquaintance with the problems that arise, can alone make him a good and efficient driver of a motor cycle, and teach him what he requires to know about the construction and maintenance of the engine and the other parts of his machine.

The Working Principle Explained.

In the four-stroke motor cycle engine, at present the most widely adopted type, a full cycle of operations is completed during two revolutions of the flywheels, commencing with the inlet (or suction) stroke of the piston, which is followed in turn by the compression, working (or explosion), and the final (or exhaust) strokes. When the No. 1 (or suction) stroke takes place, the charge, consisting of a carburetted mixture of petrol and air, is drawn into the cylinder as the piston descends, and is next forced by the following (upward) movement of the latter into the combustion chamber or compression space located in the cylinder head above the travel of the piston, where, owing to the restricted volume, it becomes highly compressed, and is then ready to be exploded with far greater force than would have been possible before compression took place.

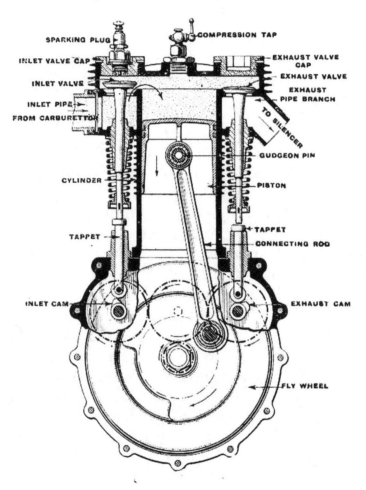

FIG. 1.—No. 1, Inlet Stroke.

The ignition of this highly compressed charge is effected by means of an electric spark, timed to occur at a precise moment in relation to the travel of the piston, but at the same time variable within certain defined and narrow limits, as will be explained later in the chapter on "Ignition." The resultant explosion acts upon the piston as a propelling force of great magnitude, causing it to descend on the No. 3 (or working) stroke, and thereby rapidly rotate the fly-wheels—to which latter, as need hardly be added, it communicates motion by means of a connecting-rod.

The next, and final, stroke of the piston impels the exploded gases upwards before it, and the exhaust valve having duly opened just before the completion of the working

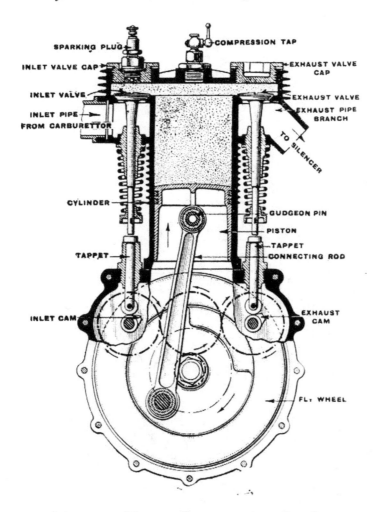

FIG. 2.—No. 2, Compression Stroke.

stroke, a way of escape from the cylinder is provided for the burnt, and now useless, gases which pass out through the exhaust pipe and silencer into the atmosphere. It will thus be seen that of the four separate strokes of the piston which go to make up the complete working cycle of the engine, only one is performed directly under the influence of the explosion, the remaining three being brought about by the momentum which is stored up in the heavy flywheels as the result of the rotative motion communicated to them during the power stroke.

Briefly summarised, then, the four-stroke cycle operates as follows :—

No. 1. The Suction Stroke↓   Charge drawn into cylinder
No. 2. The Compression Stroke ↑Charge compressed in cylinder
No. 3. The Working Stroke↓   Charge exploded in cylinder
No. 4. The Exhaust Stroke   ↑Charge expelled from cylinder

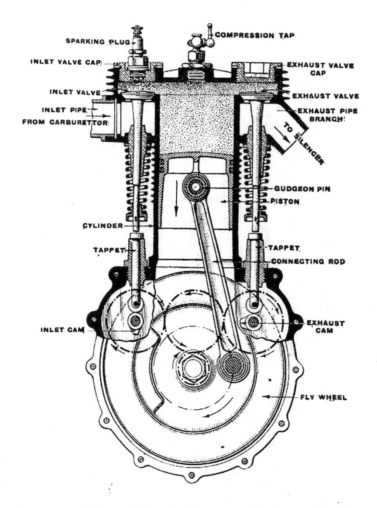

FIG. 3.—No. 3, Explosion or Working Stroke.

Reference to the accompanying diagrams (Figs. 1 to 4) will serve to further explain the description given above.

The Action of the Valves.

The inlet and exhaust valves, by means of which the gaseous charge is respectively allowed to enter and leave the cylinder, are generally of the mushroom pattern, as shown in the drawings. The exhaust valve is mechanically-operated in every case ; that is, it derives its motion from a cam, actuated from the engine main shaft by means of gear, or (as they are usually termed) timing wheels, because they govern, in conjunction with the aforesaid cam, the precise moments at which the valves—for admitting and expelling the gases to and from the cylinder—shall open and close. In the majority of cases where only one cylinder

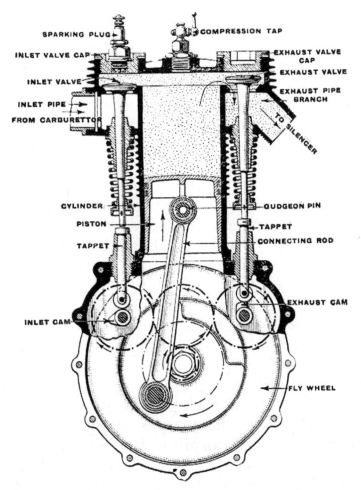

FIG. 4.—No. 4, Exhaust Stroke.

is employed the inlet valve is also mechanically operated ; but, in others, what are known as automatically-operated inlet valves (for abbreviation, A.O.I.V.) are employed, although by no means so frequently now as formerly was the case. Automatic valves depend for their opening upon the partial vacuum created within the cylinder by the descent of the piston on the suction stroke, and such valves being smaller, lighter and cheaper as a whole than mechanical valves; and, further, in view of the fact that they require no timing gear, and can be placed over the exhaust valve instead of needing a separate " pocket " (or chamber) of their own, it is claimed that they simplify the construction, and thereby reduce the chances of failure; while another advantage sometimes claimed is that the exhaust valve is effectively cooled by the admission of pure gas immediately above it.

On the other hand, as their opening movement is based upon the action of the piston, and the amount of opening largely depends upon the degree of suction induced by its speed, it will be recognised that the slower the engine runs the less will the valve opening be, so that the area for admission of the gases may be said to vary with the piston speed, and is not necessarily in accordance with the power development required from the engine. A mechanically-operated valve has a constant lift at *all* speeds, and possesses the further advantage that it is not dependent to anything like the same extent as the automatic valve upon the strength of its spring for efficient working. It seldom sticks, and with it the engine will run slower and with greater regularity on up grades.

Lack of power in a twin-cylinder motor cycle engine fitted with automatically-operated inlet valves is often traced to a difference in the strength of the inlet valve springs, and even a slight discrepancy is sufficient to affect the running of the engine, as it leads to unequal work being performed in each cylinder, instead of the driving force being equally distributed over the two pistons, as happens when both valves are correctly and uniformly adjusted. Some makers claim that for the highest speeds, and absolute freedom in running, engines equipped with A.O.I. valves are superior to those having M.O.I. valves; but if the matter is to be judged by the relative performances of each type, it must be difficult to substantiate the point.

Automatic inlet valves are undoubtedly much easier to handle than mechanical ones, but they require more constant and skilful attention to maintain them in proper condition.

Certain manufacturers of motor cycles fit their engines with what are termed " overhead " mechanical valves, such valves occupying the same position as A.O.I. valves—*i.e.*, above the exhaust valve chambers, but deriving their motion from the timing gear in the distribution chamber through the medium of long tappet rods, with an overhead rocking lever to allow of a reciprocating vertical motion induced by the action of the cam. Satisfactory results have been achieved with this system, which was first applied to track racing motor cycles and afterwards to the touring models of certain makers. The Rudge engines have always been fitted with

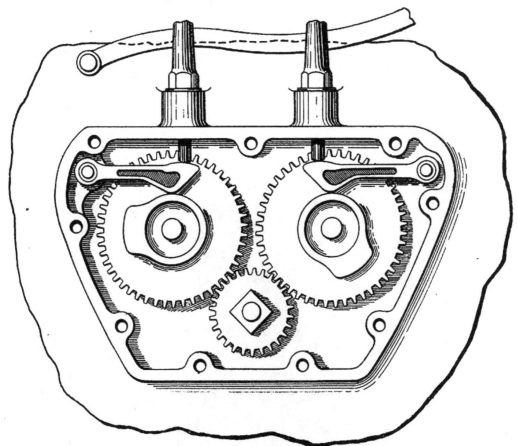

FIG. 5.—Timing Gear of a Modern Single-cylinder Motor Cycle Engine with M.O. Valves.

overhead mechanically-operated inlet valves, and this system has been retained for 1913 (see Fig. 9, page 18).

The Timing Gear.

The valve-actuating mechanism, or timing gear, for a mechanically-operated valve consists of two toothed pinions or gear-wheels, one of which is carried by the main shaft of the engine, and the other by what is known as the " half-time," or cam, shaft ; this secondary gear-wheel and shaft

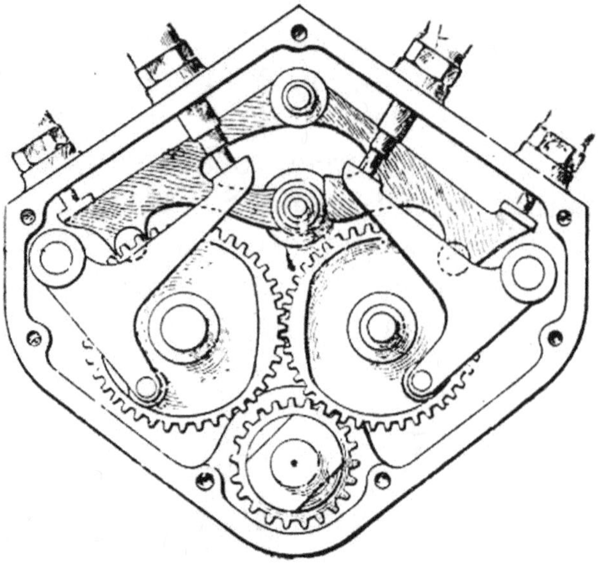

FIG. 6.—Timing Gear of
the "Chater-Lea" 8 h.-p.
Engine with M.O. Inlet
and Exhaust Valves.

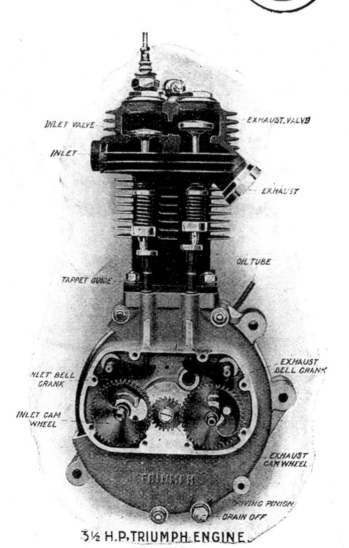

INLET VALVE

INLET

EXHAUST VALVE

EXHAUST

OIL TUBE

TAPPET GUIDE

EXHAUST
BELL CRANK

INLET BELL
CRANK

INLET CAM
WHEEL

EXHAUST
CAM WHEEL

DRIVING PINION

DRAIN OFF

3½ H.P. TRIUMPH ENGINE.

FIG. 7.—The "Triumph" 3½ h.-p. Engine, in part Sectional
Elevation.

being, as a rule, formed in one solid piece with the cam itself, and having double the number of teeth on the driving pinion, hence the use of the term " 2 to 1 " gear. As the engine shaft revolves, the gear-wheel mounted upon it, inter-meshing as it does with the cam gear-wheel, revolves the latter at half its own speed and the cam is set at the proper angle to lift the tappet, which in its turn moves the valve (whether inlet or exhaust) at certain prescribed and regular

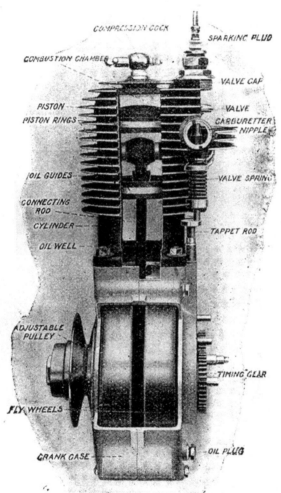

3½ H.P. TRIUMPH ENGINE

FIG. 8.—Rear End Sectional View of the " Triumph " 3½ h.-p. Engine.

periods in accordance with the movements of the piston and the stage which has been reached in the cycle of operations.

Fig. 5 illustrates the timing gear of a single-cylinder engine with M.O. inlet and exhaust valves. Matters are so organised that the inlet valve opens just after the piston has commenced the induction stroke to admit the charge into the cylinder, remaining open throughout the length of the suction

FIG. 9.—Rudge 3½ h.-p. Motor Cycle Engine with Overhead M.O. Inlet Valve.

stroke. Both valves then remain closed during the compression and working strokes, the next valve movement being the opening of the exhaust valve when the piston has reached a point somewhere about ¼ in. from the end of the working stroke, followed by its closing again as the exhaust stroke comes to an end, and the re-opening of the inlet valve, when, of course, a fresh supply of mixture is needed in order that the cycle may commence over again. Briefly stated, the cycle of movements is as follows :—

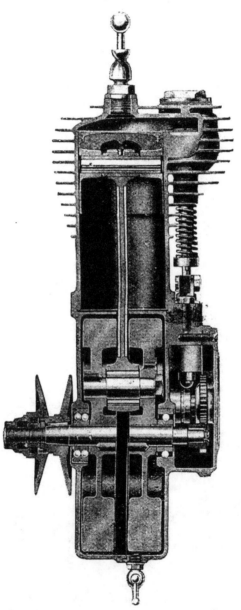

FIG. 10.—Rear End Sectional View of the " Premier " Engine.

(1) Inlet or suction stroke : Inlet valve *open*, exhaust *closed*.
(2) Compression stroke :   Both valves *closed*.
(3) Explosion stroke :   Both valves *closed*.
(4) Exhaust stroke :   Exhaust valve *open*, inlet *closed*.

The exact setting of the timing gear varies slightly in different engines, but as a general rule it follows closely upon the principle above described.

It is very important indeed that the timing of the valve movements should be correct, as only a slight error suffices to affect the performance of the engine. When once the correct valve setting has been ascertained in the builders' works, the timing wheels are (or should be) plainly punch-marked or otherwise stamped, so that in the event of its

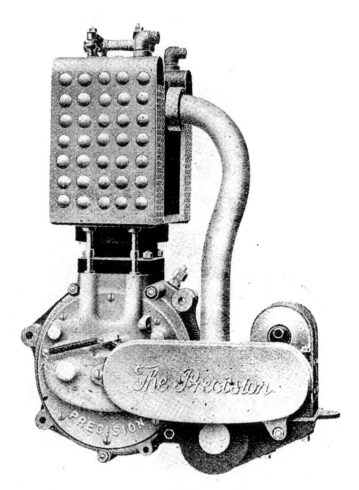

FIG. 11.—The "Green" Precision Water-cooled Engine.

becoming necessary at any subsequent time to disturb the timing, the correct relative positions of the parts may be known and no need will then arise, when reassembling them, to experiment or trouble in any way about the point.

It is customary in well-constructed modern engines to interpose between the cam and the foot of the tappet which lifts the valve a rocking arm or lever, as seen in Fig. 5

which allows of smoother action and results in a wider distribution of the effects of wear on the parts in contact, and so maintains for a much longer period than otherwise would be the case the desired amount of lift for the valve. Sometimes a roller is used in place of the rocking lever.

Springs for M.O. and A.O. Valves.

In the case of mechanically-operated valves, an average lift of about $\frac{1}{4}$ in. is considered necessary, the point depending to some extent, of course, upon such matters as the bore and stroke of the cylinder, diameter of the valve itself, port areas, and so on ; but automatic inlet valves should never be allowed more than $\frac{1}{8}$-in. opening at the most and, generally speaking, better results will be secured if this is reduced to $\frac{3}{32}$ in. The springs employed with inlet and exhaust valves must be correctly proportioned, as regards strength, for the duty they are to perform. Too strong a spring in the case of a mechanically-operated valve throws added resistance on the timing gear, and may in some circumstances result in breakage of the valve itself, due to its head being brought down on to the seating with undue force. On the other hand, too *weak* a spring, or one that has lost its life and temper as a result of long service and the effects of being heated up so frequently, makes the engine sluggish in running and liable to overheat. It is necessary that the valves should both open and close smartly (apart from the time they remain open), and that there shall be a free and uninterrupted circulation of the gases through the cylinder during the periods of induction and exhaust, and, of course, the *amount* of opening is a matter of paramount importance.

Automatic inlet valve springs are, as a matter of course, much smaller and weaker than those fitted to M.O. valves, and, in their case, it is all-important that only a correct strength of spring should be employed. Experience shows that the best plan is to use as *strong* a spring as possible in the circumstances, a normal engine of 85 mm. bore requiring a spring exercising a resistance of from 3 to 4 lbs. against the suction for full opening. It must never be forgotten that the valve will have to be opened by the force of suction against the action of the spring, so one cannot be too careful in considering the adjustments necessary in the case of an automatic valve.

Weak automatic inlet valve springs and a big opening mean sluggish running but more even firing at low speeds,

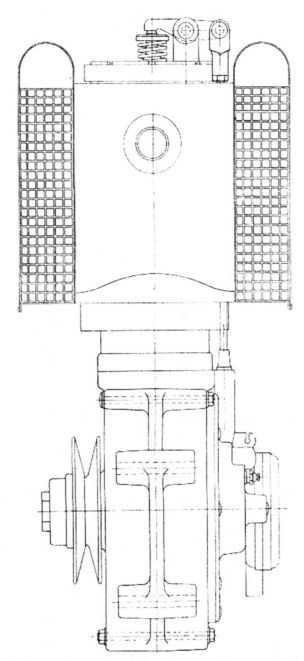

FIG. 11A.—The "Green" Precision Water-cooled Engine.

while stronger springs and a restricted opening give greater speed and greater economy in petrol consumption ; but it is not then an easy matter to keep the engine firing regularly while running slowly.

The motor bicycle illustrated in Fig. 192, page 245, is fitted with a water-cooled engine designed on the " Green " principle. The receptacle for holding the water is located in front of the petrol tank, in a separate compartment of course, but embodied with the main tank and therefore unnoticeable from outside. The water flows from there to the cylinder, alongside which latter, one on each side, are placed the radiators. The water circulates first through one radiator, then around the cylinder jacket, and afterwards through the other radiator back to the tank, and so on continuously. The placing of the radiators immediately alongside the cylinder in this way instead of as usual, separately and at some little distance therefrom, is claimed to represent a considerable advantage, and the claim, it must be admitted, is a reasonable one. Figs. 11 and 11A illustrate this engine separately.

CHAPTER II.

THE ENGINE (*continued*) : THE TWO-STROKE OR "VALVELESS" SYSTEM.

MOTOR cycle engines constructed to work on the two-stroke (or double-impulse) system have now been on the market for some little time, but their manufacture is at present restricted to only one or two firms, who have, however, made them a success, and it is more than likely that in the near future the use of such engines will rapidly develop. Objections to the type formerly included faulty carburation, excessive petrol consumption, and a tendency to overheat. The first-named has been almost, if not entirely, overcome, and the second practically so ; although it still remains the fact that two-stroke engines, as a rule, consume more petrol than a four-stroke one, the principal reason being, of course, that an explosion stroke occurs at every revolution of the flywheels instead of at every alternate one. The overheating trouble has been surmounted in one well-known design by introducing water-cooling, this design being illustrated in the present and succeeding chapters. Two-stroke engines as at present constructed do not require any distribution valves or timing gear, hence they are termed "valveless" engines, and much capital is made out of this point in their construction, which, as must be admitted, is a very good one. A valve or other device is provided for releasing the compression when starting up the engine ; but this, of course, is totally apart from anything in the nature of a gear-actuated inlet or exhaust valve.

The system of working in a two-stroke engine adapted to motor cycle practice is clearly shown in the accompanying diagrams, Fig. 12, for the use of which the author is indebted to the Scott Engineering Company, Ltd., of Bradford, who may be regarded as pioneers where the construction of such engines in a modern form is concerned. The Scott engine is of the three-ported valveless, two-stroke type, with twin cylinders and water-cooling, thus combining features which represent ideals hitherto not considered altogether feasible for motor cycle purposes.

Diagram 1 shows the piston at the commencement of the power stroke as it begins to move downwards under the force

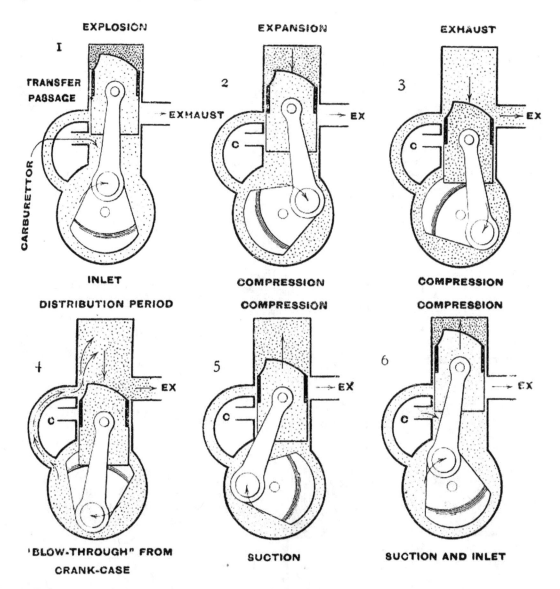

Fig. 12.—Diagrams showing Working Principle of " Scott " Two-stroke Engine.

of the explosion, whilst at the same time a carburetted mixture enters the crank-case at the inlet.

Diagram 2 shows the piston at a lower point in the same stroke, with expansion of the products of combustion in the cylinder, and the lower edge of the piston now covers up the inlet ports and thus cuts off all communication from the crank-case to the carburettor, so that the mixture begins to be compressed therein by the downward movement of the piston.

In Diagram 3 the top of the piston is seen uncovering the exhaust port, allowing the exhaust gases to escape to the

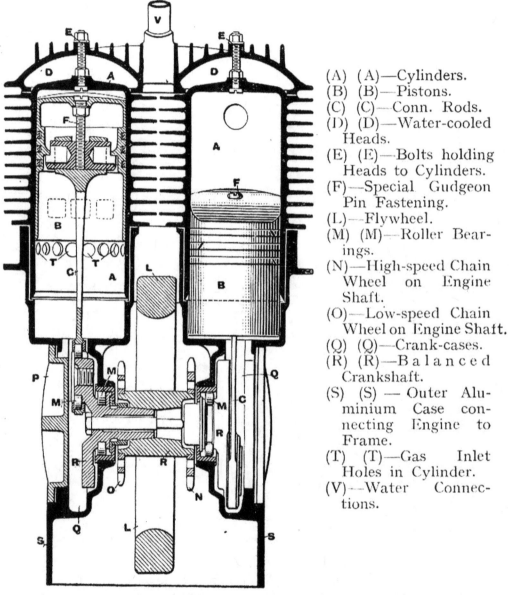

(A) (A)—Cylinders.
(B) (B)—Pistons.
(C) (C)—Conn. Rods.
(D) (D)—Water-cooled Heads.
(E) (E)—Bolts holding Heads to Cylinders.
(F)—Special Gudgeon Pin Fastening.
(L)—Flywheel.
(M) (M)—Roller Bearings.
(N)—High-speed Chain Wheel on Engine Shaft.
(O)—Low-speed Chain Wheel on Engine Shaft.
(Q) (Q)—Crank-cases.
(R) (R)—Balanced Crankshaft.
(S) (S) — Outer Aluminium Case connecting Engine to Frame.
(T) (T)—Gas Inlet Holes in Cylinder.
(V)—Water Connections.

FIG. 13.—Sectional Elevation of the "Scott" Two-stroke Twin-cylinder Water-cooled Engine.

silencer, with an instant reduction of the pressure in the cylinder. At the same time, this further downward movement of the piston causes compression of the charge in the crank-case.

Diagram 4 shows the transfer port uncovered by the piston and the compressed charge in the crank-case blows through (by way of the transfer passage) and is deflected upwards by the peculiarly shaped deflector on the piston, so that the remaining exhaust gases are swept out and the

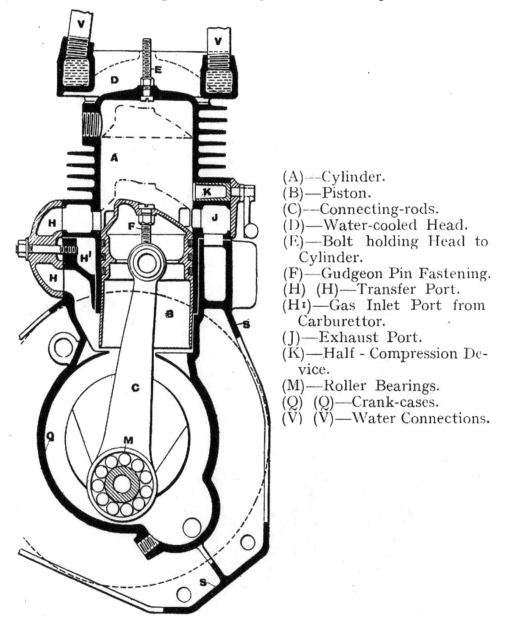

(A)—Cylinder.
(B)—Piston.
(C)—Connecting-rods.
(D)—Water-cooled Head.
(E)—Bolt holding Head to Cylinder.
(F)—Gudgeon Pin Fastening.
(H) (H)—Transfer Port.
(H1)—Gas Inlet Port from Carburettor.
(J)—Exhaust Port.
(K)—Half - Compression Device.
(M)—Roller Bearings.
(Q) (Q)—Crank-cases.
(V) (V)—Water Connections.

FIG. 14.—Cross Section through " Scott " Two-stroke Engine.

cylinder is filled with a fresh charge of carburetted mixture. This is called the distribution period of the stroke, and the efficiency of the engine depends upon the design and provision for effective distribution

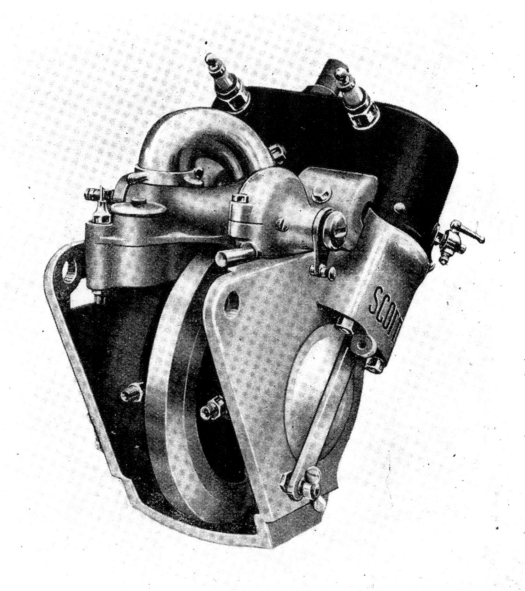

FIG. 15.—External View of the " Scott " Two-stroke Twin-cylinder
Water-cooled Engine, 3½-4 h.-p.

In Diagram 5, the piston is shown beginning to move upwards, first closing off the transfer port, whilst this upward movement creates a suction or partial vacuum in the crank-case.

Diagram 6 : The piston moves further upwards and compresses the charge in the cylinder, with a corresponding increase in the crank-case suction until the lower edge of the piston again commences to uncover the inlet ports.

At this stage in the operation, if Diagram 1 is again consulted, it may be taken as showing the piston at the end of its upward compression stroke, while by the complete uncovering of the inlet ports the crank-case suction is fully open to the carburettor and by the flow of carburetted

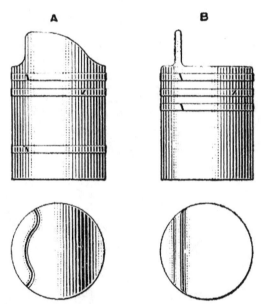

FIG. 16.—Alternative Forms of " Scott " Pistons with Deflector Ledge.

mixture therefrom the vacuum in the crank-case is destroyed, and thus a fresh charge of mixture is drawn in ready to be compressed on the downward stroke of the piston and delivered to the cylinder as described by means of Diagram 4.

It will thus be gathered that whereas in the four-cycle engine the mixture is compressed in the cylinder head by the upward movement of the piston alone, in the two-stroke engine (shown by the diagrams) it is compressed in the crank-case by the downward movement of the same, and is then allowed to pass therefrom to the cylinder, above the piston, through the transfer passage, where it is again compressed into the cylinder head before being ignited. The piston itself alternately covers and uncovers the inlet, transfer and exhaust ports, and in so doing assumes the functions of the inlet and exhaust valves as well as its own. It is formed, as will be seen, with a deflecting ledge at the top; the purpose of which is to deflect the incoming gas towards the top of the cylinder. It also prevents, as far as possible, the mingling of the fresh with the waste gases. The Scott

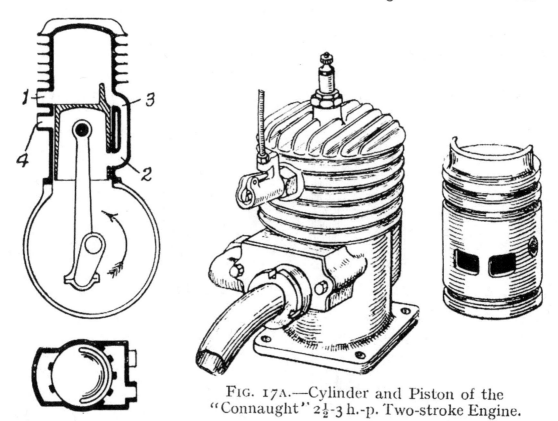

FIG. 17A.—Cylinder and Piston of the "Connaught" $2\frac{1}{2}$-3 h.-p. Two-stroke Engine.

FIG. 17.—The "Connaught" Two-stroke Engine.

Engineering Company are the originators of this system, and as designed by them the sparking plug is caused to project into the purest part of the mixture.

In the latest Scott engine, the "body" of each cylinder is water-jacketed as before, but the head is air-cooled. In a letter to the author bearing upon this point, the designer, Mr. A. A. Scott, lays stress on the fact that although the cylinder head is not water-cooled, the actual temperature cannot exceed 220° F., as is demonstrated by the fact that in the working of the engine it has never been found that the enamelling (which is stored at that temperature) is destroyed. This justifies the statement that there is practically no difference in the water-cooled and the air-cooled head, despite the fact that in commenting upon the Scott design certain writers in the motor press have criticised this feature of construction adversely.

A new design of two-stroke motor cycle for 1913 is the 2½-3 h.-p. " Connaught," made by the Bordesley Engineering Co., Ltd., of Birmingham. The engine in this case (Figs. 17 and 17A) is of the three-port, two-stroke, valveless type

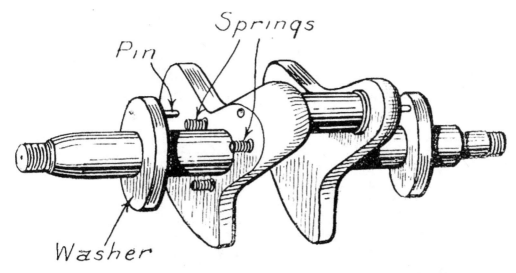

FIG. 18.—Crankshaft of the "Connaught" Two-stroke Motor.

possessing certain novel and interesting features. The cycle of operations is as follows :—

Presuming the piston to be nearing the top of its stroke, creating a partial vacuum in the crank-case, and compressing the charge in the cylinder ; the induction port is opened below the piston allowing the gas to be sucked into the crank-case. The charge on top of the piston is then fired, driving the piston down ; this closes the induction port and compresses the new charge in the crank-case. Just before reaching the bottom of its stroke, the piston uncovers two sets of ports on opposite sides of the cylinder ; one (the exhaust port) opens slightly in advance of the other (the by-pass or transfer port from the crank-case). The burnt gases rush out through the exhaust port, and the new compressed charge is forced through the by-pass port and is deflected to the top of the cylinder, and helps to drive out the remainder of the exhaust.

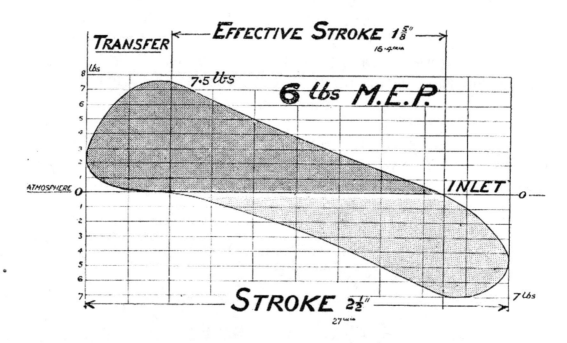

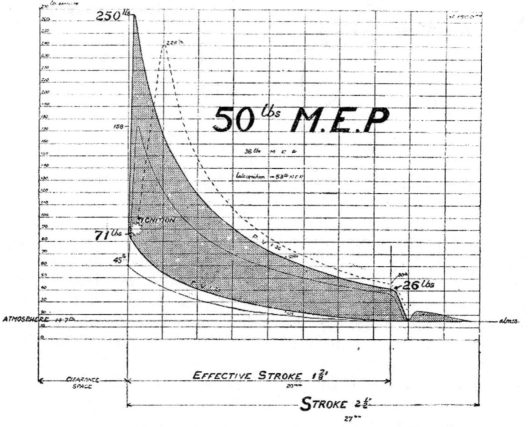

FIG. 19.—Indicator Diagrams from "Scott" Engine.

The piston then ascends, closing the by-pass and exhaust ports, and compresses the new charge and repeats the cycle.

The torque on the crank is more even, and there is less vibration in this type of engine. This (the "Connaught") motor differs in many important details from any other two-stroke motor cycle engine, and the makers claim several advantages for it. These may be summarised as follows :—

(1) The cylinder is offset forward of the crankshaft centre. This procures a more direct power impulse, reducing the side pressure of the piston against the cylinder walls. It prevents knocking, gives a maximum turning effort to the crankshaft, and increases the life of the wearing parts. It also allows a longer time for scavenging and for the new charge to enter, by reason of the slower movement of the piston at the points of exhaust and intake.

(2) The combined manifold chamber for the induction and exhaust. The induction port is on the front of the cylinder immediately under the exhaust port. By this arrangement the exhaust gases warm up the induction passage, and the incoming charge tends to cool the exhaust passage and ports, and so prevents overheating.

(3) Special design of by-pass ports through the side of piston, which prevents any oil from being thrown up into the transfer port, should any accumulate in the crank-case. It also allows a more symmetrical crank-case and cylinder, and a higher compression in the crank-case than could be obtained by carrying the transfer port down to the crank-case.

(4) No loss of crank-case compression, even though the main bearings be much worn. By means of two hardened steel collars, ground to a sliding fit on the crankshaft, but which revolve with it, and the provision of three light springs between crank disc and collar, the main bearings are completely sealed on the inside ends, no matter how much they are worn. No wear can take place between the collars and the crankshaft, as they revolve together, and any wear on the faces is at once taken up by the springs.

(5) Unique system of lubrication. This consists in simply mixing a certain proportion of ordinary air-cooled lubricating oil with the petrol. The tank is divided into two compartments as usual—one for petrol and the other for the lubricating oil ; but in place of the usual oil pump there is a special tap fitted, with a small oil measure attached. This measure

holds one-sixteenth of a pint. When filling the petrol tank,
one measure full of oil is put in with every quart of petrol ;
that is, $\frac{1}{4}$ pint of oil to the gallon of petrol. The oil dissolves
immediately in the petrol, which, on emerging from the jet
in the carburettor, is at once vaporised, and the oil enters
with the gas in the form of a very fine mist, which settles
on all the interior parts of the engine, and thoroughly lubri-
cates not only the piston and connecting-rod at both ends,
but also the main bearings, through a specially designed
collecting device in the interior of the crank-case. By this

FIG. 20.—The "Stuart" Twin-cylinder Two-stroke Water-cooled
Power Unit.

means perfect lubrication to all parts is ensured, and in direct
proportion to the amount of work done. This system is
almost universal in America with two-stroke motors, and it is
the practice in this country and abroad to mix lubricating oil
with the petrol used in aeroplane motors.

The engine is air-cooled, and has a cylinder 73 mm. bore
by 70 mm. stroke, the cubic capacity amounting to 292 cc.
The machine has already demonstrated its good qualities in
open competition on the road. The author, as a rider of the
"Connaught" motor bicycle, can testify personally to its excellent character, and as the result of his experiences with it,
can safely recommend it to those who wish for a light, handy,

and cheap running solo mount, which can also be employed with confidence for side-car work when desired.

Another interesting type of two-stroke motor cycle engine is that introduced recently by Messrs. Stuart Turner, Ltd., of Henley-on-Thames. This engine is fitted to the "Stellar" motor cycles, and has proved highly efficient in use. The engine is of the twin type, the two cylinders having a bore of 80 mm., and stroke of 82 mm., the cranks being placed at 180 degs. The engine is water-cooled, and the cylinders are so arranged that they can be worked under either the three-port or the two-port system, the latter with the "Stuart" patent inlet valve. Either system has certain advantages; owing, however, to the demand for a "valveless" engine, the three-port system is used on this engine. The crankshaft is of forged Ubas steel with special balance weights, and fitted with Stuart patent compression plates; bearings are Hoffmann balls, and fitted with Stuart patent frictionless oil and airtight sleeves; the secondary (or gear) shaft is similarly fitted. Attached to the flywheel, which is enclosed, is a well-designed plate clutch of unusually large area, operated by a lever, the spindle of which passes through an oil-tight gland. On the crankshaft beyond, and coupled to the clutch, are the gears, and still further along is the magneto drive; beyond this is another bearing, through which projects the shaft end fitted with ratchet to engage with the foot starting gear. This latter is of a quadrant gear type, i.e., a section of a gear-wheel mounted on a pin fitted to an extension cast on the gear case. This quadrant engages with a pinion mounted on a spindle housed in a small flanged casing. The spindle of the pinion and the end of the crankshaft being ratcheted, the crankshaft is subjected to a central torque strain only. Any suitable operating gear can be easily attached. Parallel to the crank-shaft is the secondary shaft. This is squared in the centre and fitted with a substantial double-ended dog, which operates for high or low gear with a neutral position. This dog is operated by a vertical spindle to which any suitable control may be attached. The gears are of large diameter, with ample width of face and suitable pitch, and are carefully hardened and bushed with phosphor-bronze. All wearing parts are of ample size—in fact, very much beyond those usually found in the motor cycle type of engine. It will be noticed that the whole clutch and gears are encased.

They are thus dustproof, and, being fitted throughout with patent bearings, no oil escapes, a point which will be greatly appreciated by riders who know the evils of oil, dust, and mud on the ordinary motor cycle engine.

The Two-stroke and Four-stroke Systems Compared.

Before leaving the subject of two-stroke engines and the point as to the increased petrol consumption associated with them in the minds of those who study these things, it may be well to draw a few comparisons in this respect between such engines and those which work upon the four-stroke principle. If we take for this purpose a twin-cylinder engine of each type having a cylinder capacity of 27 cub. ins., we find, in the first place, that owing to the fact that in the two-stroke type a certain proportion of stroke is lost owing to the piston functioning as a valve, the actual effective stroke instead of being $2\frac{1}{2}$ ins. is $1\frac{7}{8}$ ins.; hence the actual displacement of the cylinders is 20 cub. ins. instead of 27 cub. ins.; but, seeing that the charge is drawn into the crank-case, we have to compare the displacement of the crank-case with that of the cylinders in the four-stroke type.

The effective stroke in the crank-case is, in the two-stroke engine, also $1\frac{5}{8}$ ins., so that the total cubic displacement is only 20 cub. ins. instead of 27 cub. ins., and if the two engines are to be compared on the same basis, it is necessary to compare a two-stroke engine having a 20 cub. in. crank-case displacement with a four-stroke engine having the same cylinder displacement. This, in the author's view, is a very important fact, for it will be seen that the result of the reduced displacement is, of course, to make the mean effective pressure of the two-stroke engine considerably below that of the four-stroke, and whereas in the latter the M.E.P. ranges from 60 to 80 lbs. per sq. in., that in the best designed two-stroke (of course, without additional scavenger pumps) is from 40 to 50 lbs., according to the speed.

This loss is, of course, largely due to the reduced displacement arising from the less effective stroke. Another very important point is that at very high speeds the M.E.P. falls much more quickly in the case of the two-stroke than in the four-stroke type of engine, although, on the other hand, at slow speeds, the M.E.P. is increased and very nearly approaches that in the four-stroke type. In the motor cylinder itself the effective stroke is only reduced at the lower end of the cylinder. The accompanying diagrams (Fig. 19) are constructed from actual diagrams taken from a " Scott " two-stroke engine.

The whole bearing of these remarks upon petrol consumption is that it cannot reasonably be expected that the two-stroke engine should give anything like double the power of a four-stroke, although its power strokes are doubled, and the best that can be expected is that it should give double the power of an engine comparing with it on the same effective displacement.

CHAPTER III.

TWIN- AND FOUR-CYLINDER MOTOR CYCLE ENGINES (FOUR-STROKE SYSTEM).

THE object in augmenting the number of cylinders in a petrol motor is, to some great extent, although by no means entirely, that of obtaining increased power; but, in any case, whether the engine is of increased horse-power or not, when

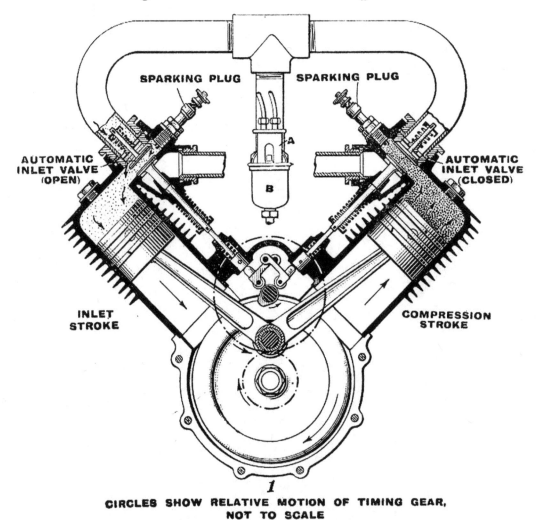

FIG. 21.—First Stage of Cycle, Twin-cylinder Four-stroke Engine.

the driving force is distributed over two or more pistons, instead of being concentrated on one alone, superior balance, lighter individual parts, and a much more efficient turning moment is secured, while starting is facilitated and there is greater flexibility and more rapid acceleration from a low speed.

In twin-cylinder engines the cylinders are usually placed so as to form a " V " longitudinally with the run of the machine, and so that they follow the contour of the frame, with the carburettor placed between the cylinders, and this

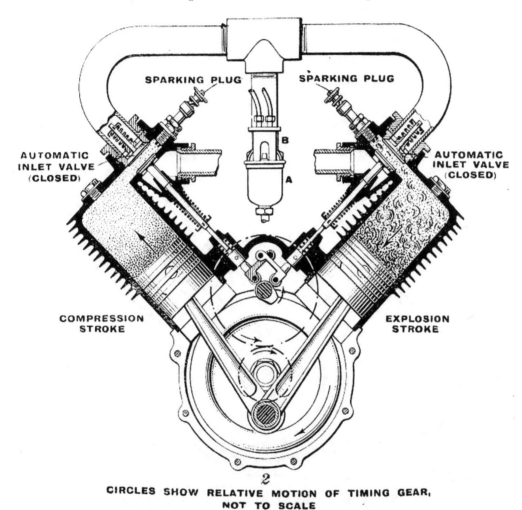

CIRCLES SHOW RELATIVE MOTION OF TIMING GEAR, NOT TO SCALE

FIG. 22.—Second Stage of Cycle, Twin-cylinder Four-stroke Engine.

forms a convenient and well-disposed arrangement, and as the crank-case remains of the same width as in a single-cylinder motor, the whole fits in place very snugly and is well adapted to the circumstances of the design. The cylinders are placed at varying angles to one another, but generally somewhere between 45 degs. and 90 degs., and the arrange-ment is usually such that when one piston is commencing its suction stroke the other is nearing the end of its compression stroke, so that ignition takes place in the latter when the former has performed some part of its suction stroke. Then, as the No. 1 piston commences to rise for compressing the charge it has drawn into the cylinder, the No. 2 piston is just about completing its working stroke, and a moment later will be travelling upwards again on the

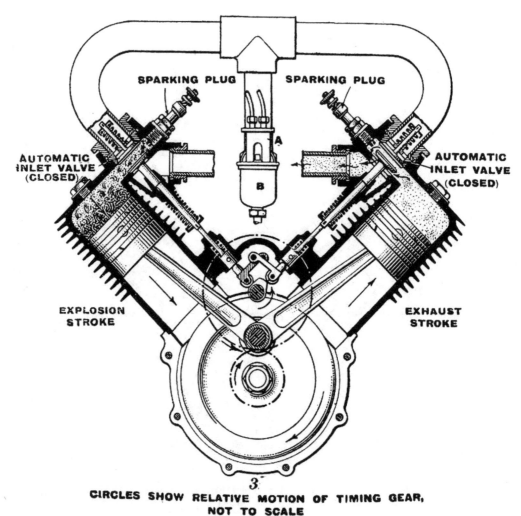

CIRCLES SHOW RELATIVE MOTION OF TIMING GEAR,
NOT TO SCALE

FIG. 23.—Third Stage of Cycle, Twin-cylinder Four-stroke Engine.

exhaust stroke; the exhaust valve having meanwhile commenced to open just before the piston began to ascend.

In the diagram (Fig. 21) on page 38, which shows an engine with its cylinders at 90 degs. apart, the piston in No. 1 cylinder is about to descend on the suction stroke, while that in No. 2 cylinder has very nearly reached the end of its compression stroke. Then in Fig. 22 No. 1 piston is rising to compress the charge, and No. 2 is practically at the end of its firing stroke, and its exhaust valve is just about to open. In this way half of the complete cycle has been performed.

In Fig. 23 ignition has just taken place in No. 1 cylinder,

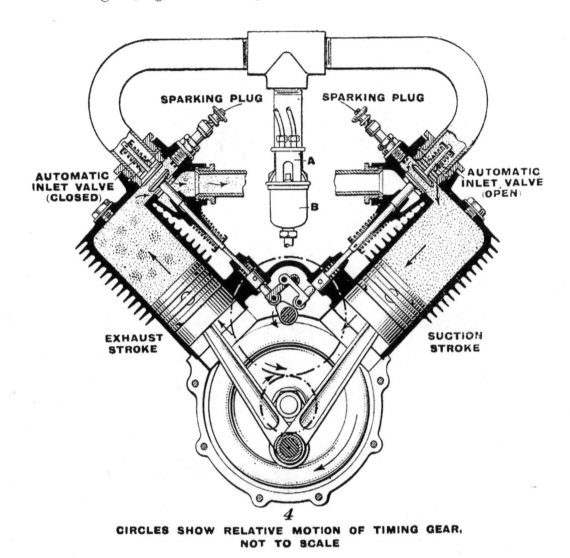

SPARKING PLUG · SPARKING PLUG

AUTOMATIC INLET VALVE (CLOSED)

A

B

AUTOMATIC INLET VALVE (OPEN)

EXHAUST STROKE

SUCTION STROKE

4

CIRCLES SHOW RELATIVE MOTION OF TIMING GEAR, NOT TO SCALE

FIG. 24.—Final Stage of Cycle, Twin-cylinder Four-stroke Engine.

and in No. 2 the piston has reached a point near the end of its exhaust stroke; while in the final diagram (Fig. 24) No. 1 piston is exhausting an exploded charge, and No. 2 piston is drawing a fresh one, and the four-cycle system being now completed, the same round of operations re-commences

Twin-cylinder engines of the " V " type are sometimes and were, formerly, very often fitted with automatic inlet valves, and the valve operating mechanism thus reduced to a single cam gear for each exhaust valve. In some engines

FIG. 25.—The " Rex " 6 h.-p. Air-cooled Engine for 1913.

one cam is alone made to do duty for both exhaust valves, so that the gear mechanism for the mechanically-operated valves is actually more simple than in most single-cylinder engines. In such designs the cam, together with the timing gear wheel, is mounted on or formed in one piece with a short projecting shaft or spindle midway between the two tappets, and actuates each rocker arm in turn. Where

mechanically-operated valves are fitted throughout, however, two cam gears at least must be employed, while in a
few cases, not perhaps of recent pattern, each of the four
valves has its own independent cam.

In a few designs of twin-cylinder motor cycle engines
the cylinders have been placed side by side transversely to
the frame. This plan was in the past successfully applied to

(A) (A)—Cylinders.
(B)—Piston.
(C)—Conn. Rod.
(D)—Crankshaft.
(E)—Inlet Port.
(F)—Inlet Valve.

G)—Exhaust Valve.
(H)—Exhaust Port.
(I) (K)—Valve Lifting
Mechanism.
(L)—Flywheel.

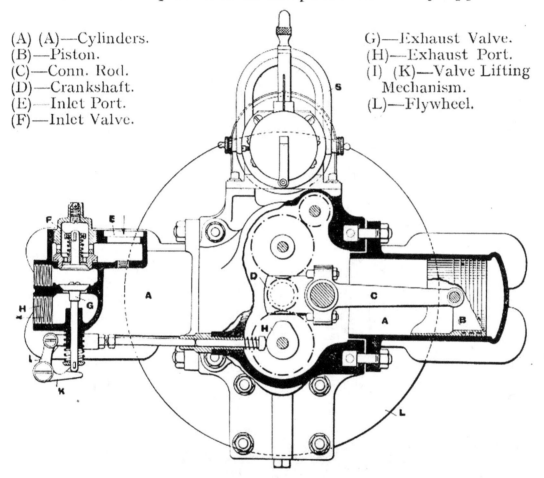

FIG. 26.—The " Douglas " Engine in Horizontal Section.

small engines for driving lightweight machines. The explosions take place alternately in each cylinder and the pistons
rise and fall in unison the one with the other. The " Scott "
two-stroke engine has its cylinders placed side by side in an
inclined position, but in this case the pistons are always
travelling in opposite directions.

In another plan the cylinders are carried in an end-to-
end position so that the pistons are horizontally opposed to
one another.

Fig. 26 illustrates an engine of a slightly earlier pattern

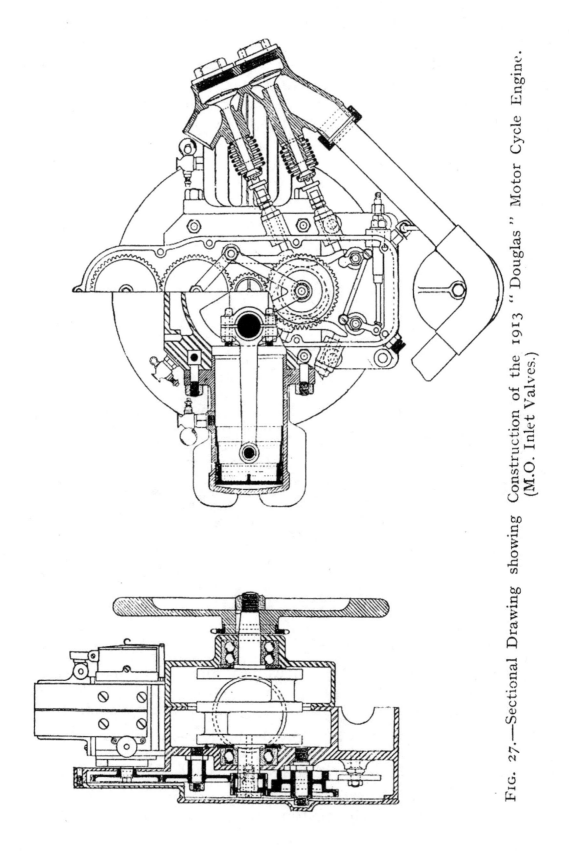

Fig. 27.—Sectional Drawing showing Construction of the 1913 "Douglas" Motor Cycle Engine. (M.O. Inlet Valves.)

than that designed for the present season. In this type the inlet valves are automatically operated and a special means of operating the exhaust valves is resorted to, the features of which may be likewise gathered from the drawing. The cam acts upon the end of a horizontal tappet rod, which in turn actuates a horizontal spindle (in reality an adjustable tappet rod) and motion is conveyed by the reciprocating movement of the latter to a bell-crank placed at its outer extremity,

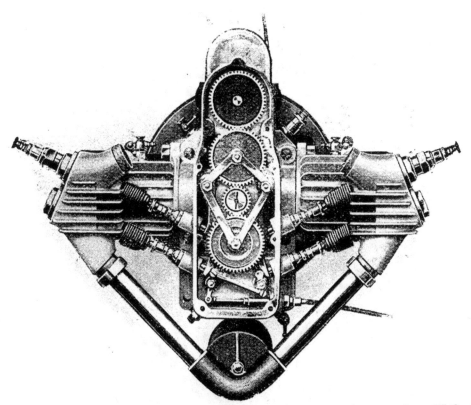

FIG. 28.—The "Douglas" Twin Engine (1913), showing Valve and Magneto Timing Mechanism.

which effects the necessary lifting of the valve itself, which is placed vertically, as shown, with the automatic inlet valve above it. In the latest Douglas engines both the inlet and exhaust valves are mechanically operated (see Fig. 27), and the arrangement of the timing gear is shown in Fig. 28. This gives greater accessibility both as regards the valves themselves and the gearing which operates them, and it is an undoubted improvement upon the previous design.

There has also been a rearrangement of the inlet and exhaust pipes and, as will be noticed, on examining the cross-sectional view, the engine main shaft runs in ball-bearings, that on the flywheel side having a double row of balls.

The magneto is mounted above the engine in the position shown, and is actuated by means of gearing encased with the timing gears. These engines have been very successful and have many good points embodied in their design.

The new Quadrant twin-cylinder engine (Fig. 29) introduced for the 1913 season is a striking looking power unit.   It

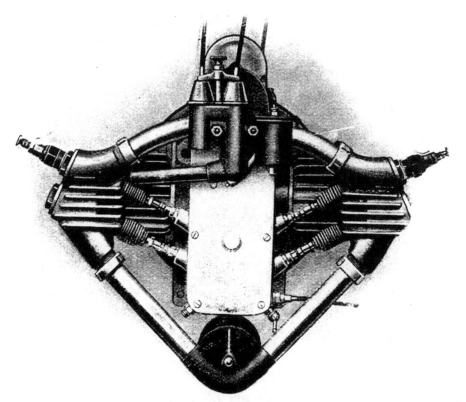

FIG. 28A.—Valve Side of 1913 " Douglas " Engine with Horizontally Opposed Cylinders.

has been designed for heavy work, especially with a side car and is well adapted to its purpose. The inlet valves are mechanically operated in the overhead position, and the timing gear for all four valves (inlet and exhaust) consists, as seen, of two double cam gear wheels driven in the ordinary manner by a common pinion on the engine shaft. The magneto gear drive consists of two pinions taking motion directly and in line from the timing wheel of the rear cylinder, there being thus only one intermediate or " idle " pinion in the whole timing system.

The valve ports in this engine are set away from the cylinders in such a manner as to give a continuous current of cold air round the whole of the port.   The cylinders each

FIG. 29.—The " Quadrant " 8–9 h.-p. Power Unit fitted with Overhead M.O. Inlet Valves.

have a bore of 87 mm. with a stroke of 95 mm. They are set at a relative angle of 50 degs. The engine, which is rated at 8–9 h.-p., is said to be both powerful and flexible, and it appears to have all the elements of good design about it.

The 6 h.-p. " Rex " twin engine for 1913 (Fig. 25) has a bore of $77\frac{1}{2}$ mm. $\times$ 95 mm. with a cubical capacity of 896 cc.

FIG. 30.—The " Precision " 8 h.-p. Twin-cylinder Engine, M.O. Valves.

The pistons are fitted with domed heads and the gudgeon pins, which are of large diameter, have a driving fit of 1-1000th taper and are locked to the pistons by means of a taper set pin and locknut. Roller bearings are fitted to the main shafts. These bearings are very easy to fit, and whilst having practically no more friction than a ball-bearing, they possess a far greater wearing surface, having a line instead of a point contact between the inner and outer races. The standard compression ratio in these engines has been adopted after careful test for the purpose of ensuring speed and power and at the same time obviating the trouble of overheating and knocking. The timing gears are of the external cam variety acting on rocking arms which lift the plungers. The engines are fitted as standard with a vapour feed from the crank-case to the cylinder heads.

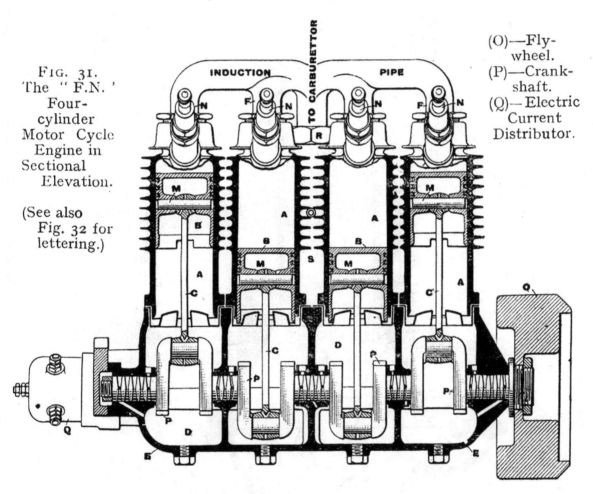

FIG. 31. The " F.N. ' Four-cylinder Motor Cycle Engine in Sectional Elevation.

(See also Fig. 32 for lettering.)

(O)—Fly-wheel.
(P)—Crank-shaft.
(Q)—Electric Current Distributor.

Four-cylinder Motor Cycle Engines.

The principal claims put forward on behalf of the four-cylinder engine are : absence of engine vibration, remarkable ease of starting, and great flexibility, combined with a positive driving effort, the latter in consequence of the fact that a shaft and bevel drive is employed instead of a belt or chain. A four-cylinder motor cycle with variable-speed gear fitted, is probably the most luxurious among single-

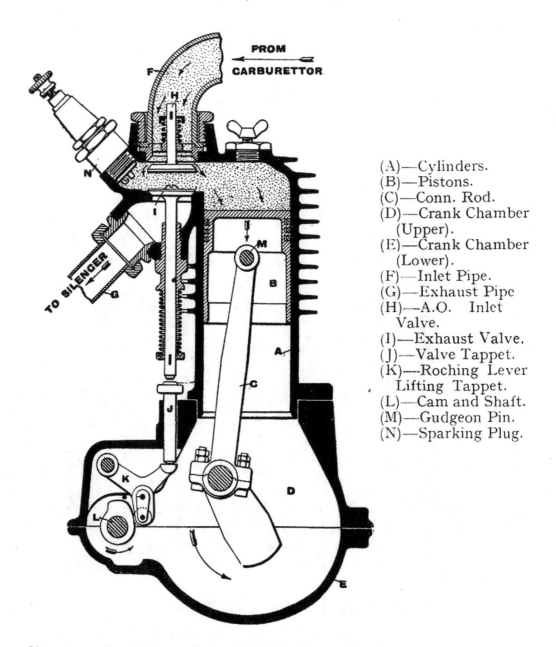

(A)—Cylinders.
(B)—Pistons.
(C)—Conn. Rod.
(D)—Crank Chamber (Upper).
(E)—Crank Chamber (Lower).
(F)—Inlet Pipe.
(G)—Exhaust Pipe
(H)—A.O. Inlet Valve.
(I)—Exhaust Valve.
(J)—Valve Tappet.
(K)—Roching Lever Lifting Tappet.
(L)—Cam and Shaft.
(M)—Gudgeon Pin.
(N)—Sparking Plug.

FIG. 32.—End View of the " F.N." Four-cylinder Engine : A.O.I.V. open ; Piston on Suction Stroke.

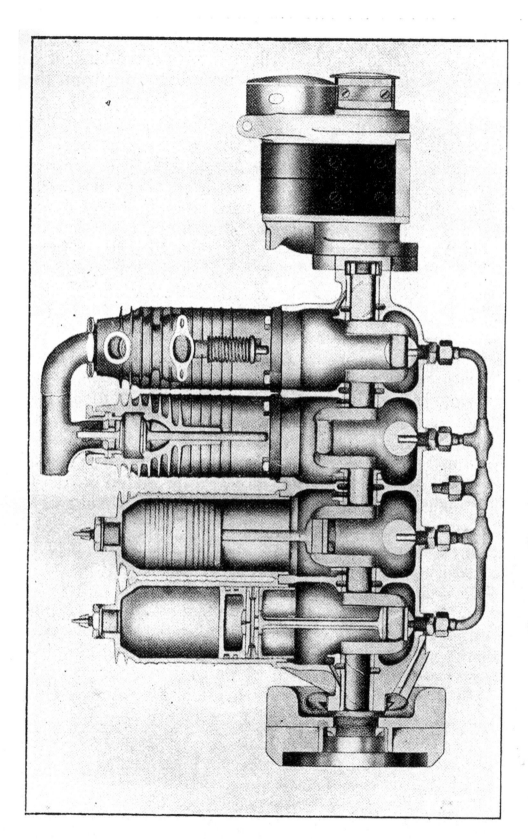

FIG. 33.—The Latest "F.N." 4-cylinder Engine.

track machines. It requires intelligent handling, and a careful eye to its maintenance ; but in return it provides a most delightful form of locomotion, and is very steady and comfortable to ride upon. Illustrated particulars of the " F.N." four-cylinder engines are appended, and from these and the other illustrations which follow on later pages a good idea of the appearance and characteristics of four-cylinder motor bicycles may be obtained.

The " F.N." engine has its four air-cooled cylinders placed in line one behind the other and mounted above a cast-iron crank-case divided into four compartments, one for each cylinder. The crankshaft is machined out of a solid drop forging and turns in five phosphor-bronze bearings. Inspection windows are provided through which the amount of engine oil in each compartment can be ascertained. Each of the four cylinders is bolted to the crank-case by three bolts and may be removed without disturbing the crank-case.

The arrangement of the cranks is shown in the longitudinal illustrations, Figs. 31 and 33. The high-tension spark from the magneto is arranged to fire the cylinders in the order 1, 3, 4, 2, counting from the rear. The flywheel is mounted immediately behind the rearmost cylinder. It contains the multiple plate clutch separately illustrated on page 140, Chapter VIII, in which chapter also are illustrated particulars of the transmission system by cardan shaft and bevel gearing, and the two-speed gear are also given.

The engine is beautifully made, and one of its principal features is the extraordinary ease with which it starts and is controlled. The smooth running and quietness of the F.N. four-cylinder motor are renowned among motor cyclists, these qualities being due to the breaking up of the power into four small units with separate and reduced impulses, whilst the positive transmission by well-lubricated, accurately fitting bevel gears does away with all jerking and irregular motion of the engine and the driven wheel.

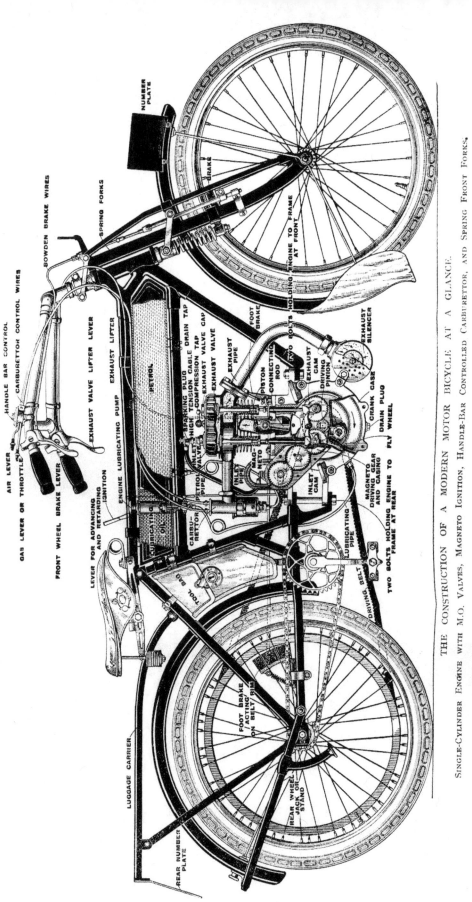

THE CONSTRUCTION OF A MODERN MOTOR BICYCLE AT A GLANCE.

Single-Cylinder Engine with M.O. Valves, Magneto Ignition, Handle-Bar Controlled Carburettor, and Spring Front Forks.

CHAPTER IV.

CONSTRUCTIONAL DETAILS OF MOTOR CYCLE ENGINES.

Cylinder and Piston.

THE cylinder of a motor cycle engine is constructed of close-grained cast-iron, and, in 90 per cent. of cases, in one piece, although a very few makers still adhere to the practice of forming the combustion head separately, with the idea of

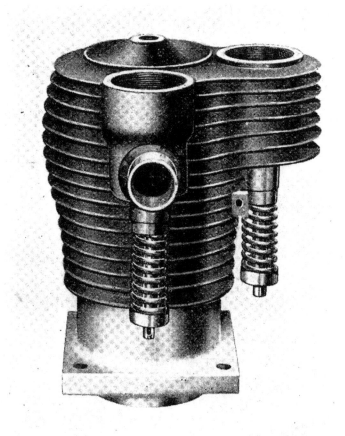

FIG. 34.—Cylinder of " Quadrant " Engine (Exhaust Valve carried in front).

providing greater access to the piston and cylinder. The cylinder is provided on the outside with a number of radiating fins or ribs, so that a largely increased surface may be available for cooling purposes as the motor passes through the air, and by this means the heat generated as the result of the continuous explosions taking place within the cylinder is dissipated to a considerable degree; hence the use of the term "air-cooled" motors.

The piston is made of cast-iron (although in a few cases of steel), and is fitted, as a rule, with three rings located in grooves near the top; but more recently it has become the fashion with some makers to employ only two rings, placed sometimes close together near the top of the piston, and at other times with one ring at the top and the other one near the base of the piston. For this last-named plan it is claimed that the movement of the piston is steadied, and that leakage of the compression is more completely obviated. It is also claimed that the use of two piston rings greatly reduces the cylinder friction and, consequently,

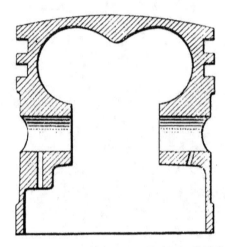

FIG. 35.—Sectional View of Piston, "Rex" Motor Cycle Engine, designed for two Piston Rings near top.

increases the engine power, and as there is naturally a smaller surface area in contact with the cylinder walls (where only two rings are employed) it seems a not unreasonable assertion. There appears to be little wrong with the three-ring method, however, when the parts are properly designed and fitted. A motor cycle recently acquired by the author has the piston fitted with two rings near the top and one near the bottom, the latter acting only as a guide, and the piston is perforated above it to assist lubrication. Some makers form the rings with square cut or "stepped" slots, and sometimes a tiny pin is inserted in the groove to keep the ring from revolving around the piston. This is undoubtedly good practice, for, should the rings work into such a position that the slots are in line, some loss of compression will almost certainly follow. In the case of the stepped-slot and pin method this, of course, becomes impossible.

Properly fitting piston rings, possessing a suitable degree of elasticity, are perfectly compression tight, and will give long service without any bother in a correctly lubricated cylinder. The top of the piston is, as a rule, flat ; but in some cases slightly domed. This latter form adds to the strength of the construction, and may also prevent, to some extent, the accumulation of burnt oil or " carbonised deposits " on the piston head, owing to the fact that the

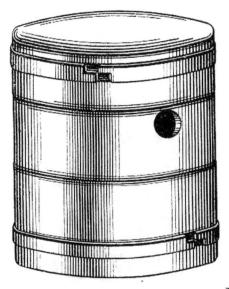

FIG. 36.

Piston with
Stepped Rings.

Piston with Stepped Rings and
Retaining Pins.

lubricating oil which has worked past the piston rings tends to drain off the domed-shaped top on to the cylinder walls instead of collecting in the centre of the piston as when the latter is flat topped.

The Connecting-Rod and Gudgeon Pin.

The connecting-rod is usually a steel stamping with bushed ends. It is secured within the piston at its upper extremity by means of a gudgeon pin running crosswise from one side to the other. The piston is provided with a short internal sleeve on each side, and the ends of the gudgeon pin are held securely in these sleeves by being driven tightly into place with a 1-1000th degree of taper, while in several different types of makers' engines setscrews are used as an additional means of securing parallel gudgeon pins. It is good practice maybe to take the additional precaution in the latter circumstances of fitting these setscrews; but the author once had a piston entirely ruined by a screw working loose and, in falling, getting fixed between the piston and the top of the crank-case, the result being a smashed piston. In another engine he drove for some 30,000 miles, and which had twin-cylinders, the gudgeon pins were held by taper

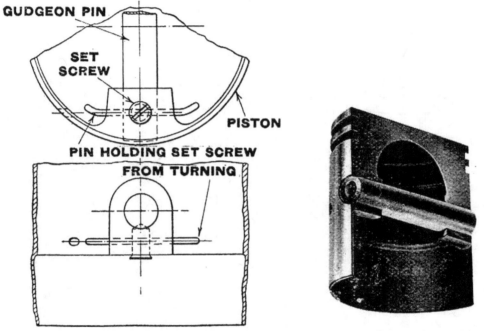

FIG. 37.—Gudgeon Pin Fastening: "J.A.P." Engines

FIG. 37A.—Piston (in section) and Gudgeon Pin: "Premier" Engine.

drive only, there was never any trouble from start to finish, and the pins never worked the slightest bit loose.

One well-known firm of motor cycle engine makers adopts a setscrew of large diameter, and further secures this by a pin passing right through the screw and the sleeve of the piston, a plan which, they say, has never been known to fail. A sketch of this arrangement is appended, Fig. 37. The upper end of the connecting-rod engages with the open portion of the gudgeon pin, between the internal sleeves of the piston; the pin itself being, of course, held stationary, while the rod swings or vibrates upon it. A gun-metal or phosphor-bronze bush takes the wear between the two surfaces. The latest " Peugeot " motor cycle engine has external recesses on each side of the piston, and the gudgeon is secured by a tapered pin and castle nut with spring washer as seen in Fig. 38.

In the majority of twin engines one of the connecting-rods is forked at the lower end, while the other rod has a single end which fits in between the arms of the fork, and both rods work on a common crank-pin as in Fig. 39.

Flywheels and their Components.

The flywheels are made as large as contingencies allow

Fig. 38.—Latest method of securing Gudgeon Pin : " Peugeot " Motor Cycle Engine.

and with heavy rims, so that full advantage may be taken of the force of momentum imparted to them during the working stroke, which motion is relied upon to keep the piston, valves, etc., acting during the remaining strokes. With heavy flywheels the engine will continue working for a much longer period after power is shut off by virtue of this stored up momentum, and will also run with regularity at a much lower speed than would be possible with lighter and smaller flywheels ; this steadiness and uniformity of running and the reduced vibration which accompanies it, is in fact a principal aim of the designer when calculating the proportions and weight of the various parts. The size and weight of the flywheels depend, of course, on the size of the cylinder and piston, the whole having to be carefully computed so that an even balancing of the reciprocating and revolving parts may be assured. The flywheels of a Quadrant engine are shown in Fig. 40 as a good example of detailed construction.

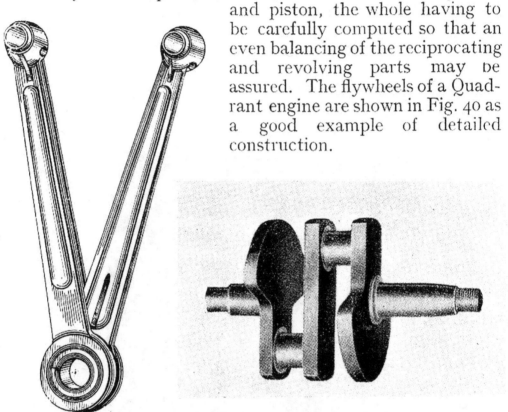

FIG. 39.—The Connecting-rods of a Twin-cylinder "V" Type Motor Cycle Engine.

FIG. 39A.—Balanced Crank Axle of the "Douglas" Motor.

The crank-pin bridges the space between the flywheels and takes the lower end of the connecting-rod in a phosphor-bronze bearing, similar to that of the gudgeon pin, but of larger dimensions. The main bearings of the engine shaft are carried in the sides of the crank-case, which are considerably thickened at these points or made with a projection on the outside so as to provide as long a bearing as possible. In some engines ball-bearings are fitted to the main shaft to further assist in overcoming frictional resistance, and thereby securing greater continuity and uniformity of rotation. Ball-bearing engines are very "lively" ones as a rule—that is, they run with remarkable freedom and dash, and the increased flexibility is especially advantageous when a lot of variation in the speed is required for traffic or other reasons.

The Crank-case and Timing Gear Chamber.

The crank-case itself is formed in two halves, and is,

FIG. 40.—Flywheels and Connecting-rod of "Quadrant" Motor Cycle Engine.

in the majority of cases, made of aluminium alloy, for the sake of lightness, but iron or steel crank-cases are by no means unknown. It is (or should be) perfectly oil-tight—that is, the joint with the cylinder, as well as that between the two halves of the case itself, should be capable of retaining oil. The joint between the base of the cylinder and the top of the crank-case can be quite efficiently made with brown paper soaked in thick oil, fish glue, or other similar composition, and in some cases such jointings are entirely absent. Some motor cyclists use Seccotine and other preparations of the same character, and then when they desire to remove the cylinder for inspection or cleaning purposes, it sometimes sticks with an obstinate tenacity, such as is highly conducive to bad temper. A thin jointing of aluminium sheet makes a good packing, but it costs more than brown paper or any of the rubberised steam packings, and sometimes cracks, and is not then of much use for the purpose. The joints between the sparking plug and the cylinder, and also those of the valve caps and compression tap, are best when consisting of copper and asbestos washers, which are both highly efficient and cheap to buy. The crank-case is fitted on its underside with a drain tap for use when running off used-up lubricating oil and when flushing out with paraffin. A non-return ball valve is sometimes screwed into the crank-case to allow of the air compressed therein escaping ; in some engines this compressed air escapes through the hollow main shaft.

The timing gears are carried in a separate chamber to themselves, known as the distribution chamber, on the outside of the crank-case, where they are easily accessible, and from which position the driving of the magneto is most readily arranged for. The timing gear wheels are of case-hardened steel with teeth cut from the solid blank, the rocking lever or roller between cam and tappet being also of steel. The cam or " half-time " gear spindle of either the inlet or exhaust valve is generally made taper at the outer end, and on to this tapered portion is secured, by means of a feather, the gear-wheel or sprocket for imparting primary motion to the magneto drive. Outside the taper again is a threaded portion so that the gear-wheel may be locked in position by means of a nut. The magneto, to which detailed reference is made in the chapter on " Ignition," has a corresponding gear-wheel or sprocket on the end of its armature shaft, and these two wheels—one on the engine half-time shaft and the other on the magneto armature shaft—are connected up either by intermediate gear-wheels or a chain.

Inlet and Exhaust Valves.

The inlet and exhaust valves when mechanically operated are of the general formation shown in Fig. 41. The stem is thickened for some distance below the head of the valve for strengthening purposes. This enlarged portion of the stem works in a guide cast with the cylinder or screwed in place below the valve chamber for steadying the valve and ensuring a true vertical movement of the same. At its lower end the valve stem has a slot cut in it for taking a key or wedge, the function of which is to hold the cap in which the spring rests at its base and prevent the spring from falling. The spring, as already shown, exercises a good deal of control over the movement of the valve after it has been lifted from its seating by the cam, and, when of accurate strength, ensures a prompt and even return of the valve to its seating. The cam, as before pointed out, has to operate against the power of this spring; but without the latter there would be no certainty or regularity in the valve movements.

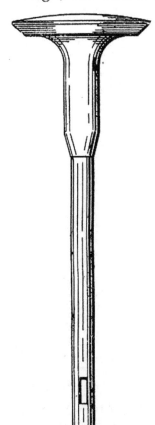

FIG. 41.
Mechanically operated Valve of Motor Cycle Engine.

The cam acts on a tappet rod, which in turn raises the valve, and it is very important that the tappet should be a true and easy (but not slack) fit in its guide, so that it may move quite freely but yet not allow oil to escape from the crank-case through the guide. The tappet rod usually has a disc or " table " head, which tends to distribute the wear between it and the foot of the valve stem instead of such wear being concentrated on one fixed point all the time; it also ensures that an even contact between the two parts will permanently maintained.

It is even more important still that the valve stem be an accurate fit in their guides, as, owing to the considerable heat which is imparted to them, expansion takes place, and unless this is duly allowed for (it sometimes is *not*), the valves will stick in their guides and then there is an end for the time being to the proper working of the engine.

As before said, the valve springs must be correctly proportioned to suit the particular engine to which they are applied, and when it becomes necessary, as in course of time it must do, to replace them, the best plan by far is to obtain the new springs direct from the makers of the engine rather than to select them haphazard on the "nearly as possible the same" principle. The difference in the performance of an engine when the springs are of proper strength and in good condition, and when the opposite is the case, is a very appreciable one indeed; and often, when the rider has exhausted all his endeavours in trying to find out the cause of unsatisfactory running, a simple replacement of the valve springs will work wonders, immediately restoring the lost power and "life" of the engine. Mechanically-operated valves are, as a rule, made in one piece, of nickel steel, with the heads as large as circumstances permit, so that the free passage of the gases may be assured, without its being necessary to provide a big lift. The stem is also of substantial proportions in good class engines, and the whole is turned out in accurately finished style, which is more than could have been said a few years ago where certain makes were concerned. Some makers fit their valves with cast-iron heads, and others with cast-iron rings to form seatings.

Automatic inlet valves work with the head downwards and the short stem pointing vertically above them, as illustrated in Fig. 32. In this type the seating is removable with the valve itself, this greatly facilitating the adjustment or grinding of the valve, the latter operation being very rarely needed, but the former one frequently so. The guide for the stem is in one with the seating, and the latter screws into the combustion chamber in the top of the cylinder above the exhaust valve pocket. The spring and stem are retained in their correct relative positions by various means, either a cup and cotter being used much as in mechanically-operated valves, or else the valve stem is screw-threaded and the parts are secured by means of locknuts. The method of fixing and adjusting the valves is described later in Chapter XI, which deals with "Overhauling and Tuning-up the Engine."

The inlet valve chamber is in communication with the induction pipe, to the other end of which the carburettor is attached, whilst the exhaust valve chamber has a branch to which the exhaust pipe is attached, and by means of which the burnt products of combustion pass away to the silencer and finally into the atmosphere.

Cylinder Mountings.

An opening is formed in the combustion head of the cylinder, and this is screwed to receive a compression tap to allow of the " compression " being released if desired, and also for injecting petrol or paraffin into the cylinder to facilitate starting the engine in cold weather, should the piston have become somewhat " gummed " up by the film

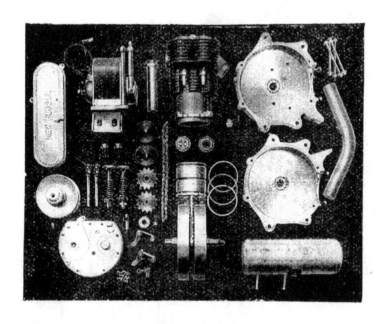

Fig. 42.—A Motor Cycle Engine Dissected : The " New Hudson " $3\frac{1}{2}$–4 h.-p.

of congealed lubricating oil which sets between itself and the cylinder walls. This tap is seldom used for the purpose of releasing the compression, that operation being much more conveniently achieved by means of the exhaust-valve lifter, a device which permits of the exhaust valve being lifted from its seating altogether independently of the cam, so that the compression escapes down the exhaust pipe until the valve is dropped on to its seating by the release of the lifter handle. A second screwed opening in the cylinder head takes the threaded shank of the sparking plug, communication between which and the accumulator or magneto is established by means of a high-tension cable for the conveyance of current for igniting the charge by means of an electric spark, the whole of these details being either separately or collectively illustrated in one portion or another of these pages.

Exhaust Pipe, Silencer, and Cut-out.

The exhaust pipe communicates, at its upper extremity, with the exhaust valve chamber and delivers the burnt

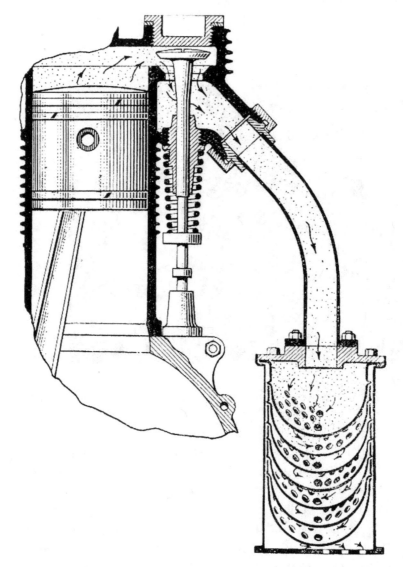

FIG. 43.—Path of Exhaust gases through "Clair" Silencer.

gases to the silencer below, the passage of which into the atmosphere is retarded somewhat by means of baffle-plates, cups, or other means of breaking up the discharged volume into particles before allowing of final escape. Very much

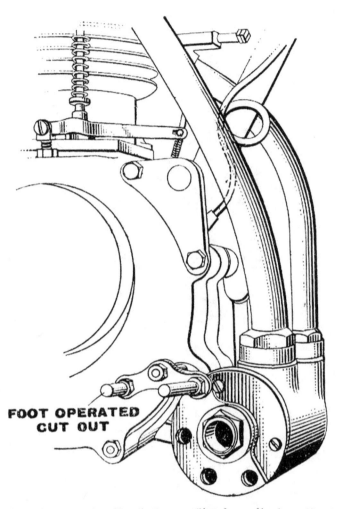

FIG. 44.—The "Premier" 3½ h.-p. Single-cylinder Engine with Auxiliary Exhaust Port and two Exhaust Pipes.

This arrangement permits of the hottest gas being released immediately it has done its work, and so enables the engine to keep cooler. It also reduces internal friction by relieving the pressure against the piston on the exhaust stroke and increases the life of the main exhaust valve. The auxiliary exhaust valve is located in the front of the cylinder near the bottom of the stroke, and a light spring keeps it on its seating while fresh gas is in the cylinder. It is adjustable by means of a setscrew and locknut.

5

indeed depends upon the design of the exhaust pipe (which should be of ample area and free from sharp bends), and more still upon the design of the silencer itself ; for if the opposition to the free egress of the exhaust gases is too strict, the latter, so to speak, become trapped and are unable to get clear with sufficient rapidity, with the result that back pressure is set up ; the engine overheats, and loses power.

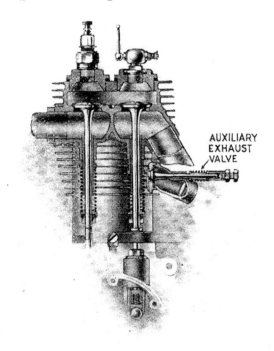

AUXILIARY EXHAUST VALVE

FIG. 44A.—" Premier " Engine Cylinder with Auxiliary Exhaust Valve.

A reasonable degree of silence is highly to be desired, but if it be accompanied with anything more than a slight amount of back pressure, then it is dearly bought. The accompanying drawing (Fig. 43) shows one of the well-known " Clair " silencers, the efficiency of which has now been fully established. It will be seen that the silencer chamber contains a number of cups, each having a number of small holes, and the path of the gases is clearly shown from the cylinder to the atmosphere. In the author's experience the use of the " Clair " silencer gives a remarkable degree of silence, devoid of back pressure altogether. The gases are afforded a free exit without being allowed to escape until well broken up, and then at a low velocity.

Some " silencers " are unworthy of the name ; but others, while being of very simple construction, give quite good results. The addition of a length of piping with flattened end to the silencer sometimes aids matters, and the escape holes are not then required. Many manufacturers of motor cycles equip the silencer with a " cut-out," which may be opened and closed by the rider at will. When opened the escape of the gases is largely facilitated, and there is a corresponding increase in the noise occasioned, but the engine usually accelerates, and will, as a rule, climb a hill better when the cut-out is open than when it is closed. The indiscriminate use of the exhaust cut-out when passing through towns, or, indeed, at any time other than when riding on an open country road, is to be largely deprecated. In the case of twin-cylinder motor cycle engines a separate silencer may be used for each cylinder, or the two exhaust pipes may be led into one common silencer. Provided that the correct volume of space is assured, it, perhaps, matters little whether one or two silencers are used. The single · silencer plan facilitates the fitting of an exhaust cut-out on a twin engine.

Since the above was written, the Local Government Board has introduced new regulations prohibiting the use of exhaust cut-out appliances such as are fitted between the cylinder and the silencer. The exhaust gas must pass first through an expansion chamber or silencer before being allowed to escape into the open air. In spite of this regulation long exhaust pipes with flattened ends are, it appears, permissible, provided that the requirements as to noise are met thereby.

CHAPTER V.

COMPRESSION.

HAVING now considered the working principle of the engine—both in its four-stroke and two-stroke forms—and also the general features of design and construction, it will be useful to turn our attention to the all-important subject of Compression, upon which phase of its being the petrol engine depends so largely for its efficiency. We have already seen that, after the gaseous mixture from the carburettor has been drawn into the cylinder on the suction stroke, the following upward stroke of the piston compresses it into the restricted space known as the combustion chamber at the top of the cylinder, so that, as before explained, the explosion may be greatly magnified in force, and it will be readily understood that, unless every part of the cylinder and its components be made gas-tight, the compressed gases will leak through and be lost in part for the purpose of driving the engine.

Loss of compression occurs when the piston rings do not spring out all round to tightly fit the cylinder walls, or when the valves, especially the exhaust, does not bed down properly in its seating. A very hot crank-case is a sure indication that the piston rings are a bad fit, and allowing compressed gas or air to leak past them. The joints between cylinder and the compression tap, sparking plug, and valve caps must all be securely made to resist leakage, and, of course, there must be no crack of any sort in the cylinder casting extending to the interior.

In course of time, when the engine has done a considerable amount of work, the exhaust valve face becomes worn and " pitted " with small indentations, resulting from the contact of the valve with burnt gases at a very high temperature, and when this stage is reached, it becomes necessary to grind the valve carefully in its seating until a smooth and even surface between the two has been restored—a very easy, if somewhat laborious and tedious, process. Once done—and carefully done—further loss of compression in this direction is, for a considerable period of working, only prevented, and if not then entirely satisfactory, other causes must be looked for in the directions indicated above.

When all the points at which compression can leak away are made sound and in good order, the piston rings springing out and forming a true contact all round with the cylinder walls, with the gap in the rings not more than 1-64th in., valves an accurate fit in their seatings, and all the other parts well conditioned, then the engine is sure to have a good compression, and one which should withstand the weight of the rider suspended on the pedal while the rear wheel is jacked up off the ground for at least 30 seconds, or, when pushed along the road, skid back the wheel as the highest point of compression is reached. With the compression in good order, the engine will start more readily, and will perform in better style on hills. Even a slight loss of compression will affect its working, although, maybe, not very noticeably ; but if the leakage be at all serious, it may become a difficult matter to get any really useful work from the engine, except under the most easy and favourable circumstances. Leakage of compression is more serious at low than at high speeds, as then the power developed by the engine is in any case less than when running fast.

The Ratio of Compression.

In addition to the question of good or bad compression there also arises that of *high* and *low* compression—another matter altogether. A high-compression engine is one in which the cylinder combustion chamber or compression space is restricted in volume more than is usual, so that the gases are more tightly compressed than where the volume is larger. The ratio of compression depends very largely upon the purpose for which the engine is required, and the degree is also affected by valve areas and timings, combustion chamber design, and flywheel diameter and weight. For very smooth and quiet running a compression ratio as low as 35 lbs. has given very good results, although, with this compression, it becomes necessary to ignite the charge much earlier than is usual. For all-round purposes a ratio of 3 to 1—that is, cylinder capacity three times that of the combustion chamber—has proved very satisfactory.

With this compression fairly high speeds may be obtained, and, generally speaking, the engine will plug up a long hill, as well as potter along very slowly if required to do so. This ratio is a much favoured one with motor cycle manufacturers for touring machines, and the majority of them probably work within a few pounds of it. There is, no doubt, some advantage to be gained in the direction of economy and speed by using an exceptionally high compression if the aim is short-distance bursts of speed only, as in competition work, for example; but for continued running the heat generated is such as to absolutely prohibit any continuation of running with a wide throttle, if the compression is anything over 90 lbs. Above this ratio the mean working pressure does not increase in proportion to the amount of compression, and a great deal of negative work is thrown on to the engine.

A compression of 90 lbs. in an air-cooled engine makes it impossible for a well-warmed engine to take full advantage of throttle opening, as the expansion caused by the heat generated by the work done in compressing the charge brings the pressure so high as to nearly approximate to that due to the explosion; in fact, the charge is frequently ignited previous to the piston completing the compression stroke, and what is called " knocking " then takes place. The burning of the charge under these circumstances may almost be likened to detonation, as compared with the expansive push really required.

The effect of compression alone in increasing the rate of combustion is due to the fact that particles of the various gases of which the explosive mixture is composed are then in much closer intimacy. This is not the only advantage of compression in a petrol engine. Other advantages are that, with a given engine stroke, there is a large change of volume between maximum compression and the end of the stroke. Also that, by compressing the gas, a large weight of gas is contained in a small volume, so that the distance through which combustion has to extend is small, and the amount of cooling surface retarding combustion is also small. The degree to which it is advisable to carry compression is to some extent limited by fear of producing a jerky engine, but chiefly by the possibility of self-ignition of the gas before the proper moment, owing to the heat produced by compression.

CHAPTER VI.

THE CARBURETTOR.

THE carburettor (or vaporiser), although not exactly a part of the engine itself, occupies a position of the greatest importance in relation thereto; indeed, it may be said that the efficiency of the engine (assuming the other parts to be in order) depends primarily upon this relatively small device, the function of which is to feed the mixture or fuel, for consumption in the engine, in the right proportions as regards its constituent elements and to maintain a continuous and sufficient supply.

In other words, the carburettor is a device in which the petrol delivered from the tank is turned into a highly gaseous vapour by its mixture with air, and, unless the mixing process is performed with accuracy and in the right proportions, faulty working of the engine follows as a matter of course. In general outward and detailed design different makes of carburettors vary considerably; but the main operative principle remains the same in all cases, so that a brief description of the principle of carburation, in conjunction with some examples from present-day practice in the design and construction of the apparatus, will suffice to make clear what the carburettor is and how its functions are performed.

Spray and Surface Carburettors.

The modern carburettor, as applied to motor cycle engines, is invariably of what is known as the "spray" type—that is to say, the petrol is delivered to the chamber in which the mixture with air is to be effected in the form of spray or a stream of small particles; whereas, in the old surface carburettor, formerly always used, the petrol collected within a chamber in a pool and the current of air was so arranged as to be drawn over its surface and so to set up the process of vaporisation.

Some of these carburettors had wicks for soaking up the petrol, thus increasing the surface exposed to the air, and there may also have been other methods designed to effect

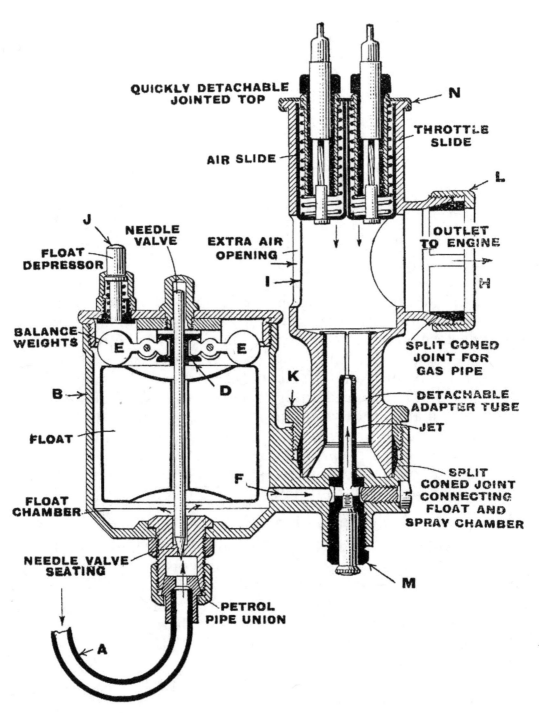

FIG. 45.—The " B. & B." Carburettor in Sectional Elevation.

the same purpose. Whatever they were, however, they have now all given way to the spray form of construction, which, although perhaps necessitating a rather more complicated apparatus for its accomplishment, takes up less room and is doubtless better suited to modern requirements.

Construction and Working of the Spray Type.

The spray carburettor consists in the main of two separate portions—*i.e.*, the float chamber, and the mixing or spraying chamber. Petrol from the tank passes down through the petrol pipe (A), Fig. 45, to the float chamber (B).

In this chamber, as its name implies, there is a float which consists of a hollow brass body or drum, through the centre of which there passes in a vertical direction a long thin rod, known as the needle valve. The base of this needle is tapered to a point, and this conical portion fits into a tapered opening through which the petrol rises into the float chamber. The function of the needle valve and float is to control the feed of petrol from the pipe communicating with the main tank, and the manner in which this function is performed may be briefly described as follows :—

Near its upper extremity the needle has affixed to it a grooved collar (D) with which there engages a pair of small brass balance weights (EE), each pivoting on a small split pin and held securely in position by a bracket through which the split pins also pass, and which is itself secured to the cover or lid of the float chamber. The needle rests at the top in a tubular sleeve raised above the cover and, as seen, clearance is left above the needle so that the latter has space in which to rise. When the petrol reaches the base of the float chamber it passes directly into the chamber itself, causing the float to rise and to continue doing so until the two brass weights (EE), which are pushed upwards by the ascending float, are raised to a predetermined height, at which point they automatically cut off further supply by forcing the needle down on to its seating and the further delivery of petrol is for the time being cut off. The passage (F) forms a means of communication between the float chamber and spraying chamber. Here the petrol enters what is known as the jet, or spraying nozzle, a thin vertical column of brass having a vertical central passage of very small diameter through which a fine stream of petrol rises and finally issues from the orifice in the form (as before said) of spray, to be mixed with air, and thus converted into the vapour known as petrol gas. The jet is screwed into position from the underside and is adapted, in the latest carburettors, so that it may be readily removed, or various other sizes substituted, without interfering with any other part.

The spraying chamber is provided underneath with apertures, through which a permanent supply of air can enter for mixing with the petrol spray as it rises from the jet orifice. This is called, for the sake of abbreviation the " fixed " air intake.

It is obvious, however, that the main parts of the carburettor cannot be changed to suit different types and sizes of engines, so a little device known as the adapter, or choketube, is employed. The function of this is to shut off a certain percentage of the permanent air supply, and in this simple manner it is possible in a very short time to find the exact requirements of any particular engine.

The bore of the jet and the diameter of the adapter are calculated on the basis of the size and power of the engine to be fed. If the engine is large, a large size adapter is used, with a correspondingly large jet, the sizes varying proportionately. The adapter merely consists either of a piece of plain tubing formed of thin brass with a small flange at the top ; or it may have a truncated portion at the bottom to act as a funnel or intake for the air. In addition to the fixed air supply, means are provided for admitting to the spraying chamber a variable supplementary supply, this being called the " extra " air, and controlled by a valve or slide which may be adjusted to give any degree of opening that the circumstances call for by means of a small handle provided near the rider's hand for the purpose.

Carburation and Successful Working.

The whole secret of successful working in a petrol engine is to get the " mixture," as the gaseous fuel is generally called, correctly proportioned in respect of the percentages of petrol and air which combine to produce it. A weak mixture—*i.e.*, one into which air enters too largely, causes weak explosions, owing to the fact that the gas supplied to the cylinder lacks the substance required to produce the necessary force on being ignited, so that power is lost in driving the piston and the engine fails to perform the task required of it ; and if the mixture is *very* weak, it may be quite impossible to obtain explosion at all. On the other hand, if the gas is too rich—that is, without its proper admixture of air—a similar result will follow due to imperfect combustion, and, in any case, the rate of petrol consumption will be increased and overheating of the engine may ensue.

When the fuel delivered to the engine is correctly mixed, there is a wonderful difference in the working of the latter, and every motor cyclist should endeavour as much as possible, both in his own interests and that of the engine itself, to study this very important aspect of manipulating a petrol engine. The further remarks to be found in Chapter XII, which deals in part with the subject of driving a motor cycle, will, it is hoped, assist the reader in this direction.

The space above the jet forms a chamber in which the gas forms, and from whence it passes directly to the engine cylinder. Its upper cylindrical portion contains the air and throttle valves, or slides, by means of which the admission of the gas and the " extra " air are separately controlled. These valves are now-a-days operated from the handle-bar by means of Bowden cables, hence the term " handle-bar controlled " carburettor. They were formerly operated by small handles, attached to the sides of the petrol tank, which actuated rod and lever mechanism communicating with the carburettor, and this system one hears referred to as " tank control." It is now obsolete so far as modern construction is concerned.

Each valve or slide consists of a hollow body, usually of semi-circular or D-shaped section, containing a spring which assists in controlling the movements of the valve. The outlet branch-piece (H), by means of which the carburettor is coupled up to the induction pipe of the engine, communicates with the interior of the spraying and valve chamber, which latter at its opposite side has an opening (I) for the admission of the supplementary air supply, this opening being controlled by the air valve operated by the rider (as before said) from the handle-bar. The float chamber is provided with a plunger (J) by means of which the float may be depressed, and the needle valve consequently raised and kept off its seating, so that the float is prevented from cutting off the petrol until the chamber is full.

This is what is known as "flooding" the carburettor, and it is usually resorted to prior to a start, so that the petrol may rise in the jet and the action of the carburettor be in part commenced while the engine is as yet stationary.

When the engine is working the petrol is drawn through the jet by the suction stroke of the piston, further assisted, maybe, by the current of air passing upward through the permanent openings in the base of the mixing chamber ; and when the engine in the course of its working has depleted the supply in the float chamber, the float gradually descends and the needle valve opens to admit a fresh supply.

Thus we have the following system of action :—

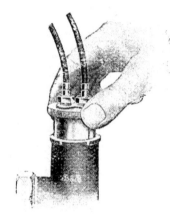

FIG. 46.—Removing Jet from " B. & B." Carburettor.   FIG. 47. — Removing Air and Throttle Slides, " B. & B." Carburettor.

1. Petrol flows from tank through petrol pipe to the float chamber.

2. Petrol enters the float chamber and raises the float until the balance weights reach a predetermined height at which point the needle valve is forced back into its seating and cuts off further ingress of petrol until the float is again allowed to descend as the engine consumes the petrol which is keeping it up. As the float descends so the balance weights also move with it, and the needle valve is again lifted to let in a further supply of petrol.

3. Petrol passes through small passage into the jet, and, so long as engine is working, is drawn through it by the suction of the engine, and being mixed with air, passes from the carburettor through the inlet pipe to the inlet valve chamber of engine and thence into the cylinder, in the form of a gaseous vapour of a highly explosive character.

The " B. & B." Carburettor.

The carburettor selected to illustrate the principle of carburation, as above described, is that known as the " B. & B.," manufactured by Messrs. Brown & Barlow, Ltd., of Birmingham. Although not of the makers' latest pattern, this design is in use on many hundreds of motor cycles, and the drawing lends itself especially to the purpose of explaining the general principle on which this type of carburettor operates, and for this reason it is retained in the present edition. The author's long acquaintance with this make of carburettor has led him to form a very high opinion of its efficiency, and, in the latest patterns especially, everything possible has been done to ensure instant accessibility to every part, while the general construction is of the most simple character and the workmanship throughout of the highest class. The sectional view on page 72 shows the general construction of the carburettor. The needle valve is completely covered in and its seating is made separate from the body, and can therefore be readily replaced if desired. The spraying chamber is fixed to the float chamber by a special form of clip, the same arrangement being used for attaching the carburettor to the inlet pipe. By this method—to detach the float chamber from the spraying chamber—it is only requisite to unscrew the nut (K)—or, in the case of the later patterns, the pin—half a turn, when the spraying chamber can be turned in any direction in relation to the float chamber, or completely withdrawn for inspection if desired. The same applies, as before said, to the fastening between the carburettor and the inlet pipe. Half a turn of the union nut (L) releases the carburettor, when it can be at once detached.

The jet is detached from underneath by simply unscrewing the small nut (M), Fig. 45, when the jet can be withdrawn in less time than it takes to write it. The throttle and air valves, with all their parts, are made under Messrs. Brown & Barlow's patented bayonet joint, which renders them extremely easy of access. A quarter turn of the cap (N) allows the cap, complete with adjusting bushes, air and throttle valves, and all their parts to be withdrawn for inspection, when the adapter can be removed and, if desired, replaced by another one differing from it in size. Any one of the above operations can be effected without interfering with other parts of the carburettor.

Messrs. Brown & Barlow, Ltd., have also introduced an improved type of carburettor. In this design (Fig. 48) the jet is of the adjustable pattern, and the main air supply through the base of the spraying chamber is also adjustable. The jet opening may be varied from zero to ·036 by means of the small outstanding lever seen in the illustrations at the bottom of the spraying chamber. This lever actuates a plug or column having at its upper extremity a mushroom-shaped head which moves eccentrically over two holes, thus varying

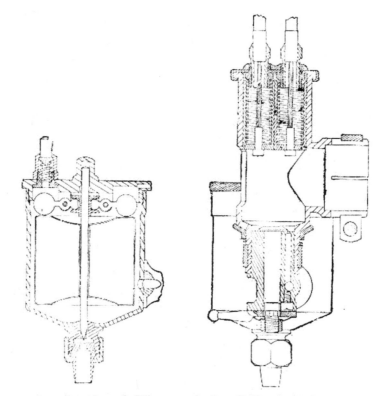

FIG. 48.—Sectional Views of the " B. & B." Universal Carburettor.

the jet opening by gradually covering or uncovering them. The area of the two apertures combined equals a jet opening of ·051.

A series of holes is formed in the base of the spraying chamber, these being opened and closed by means of a rotatable slide which allows the number of holes open to vary from one to eight, thus admitting a smaller or larger volume of air as may be desired.

The air and throttle slides are now so made as to render it impossible to insert them in the wrong positions as was possible in the earlier patterns. The jet holes are rendered immediately accessible by removing the small screw seen in Fig. 48 alongside the vertical column with eccentric head previously referred to. A small pocket spanner is provided, by the aid of which the whole carburettor can be taken to pieces without the use of any other tool. This carburettor is an undoubted improvement upon its predecessors.

For the 1913 season Messrs. Brown & Barlow, Ltd., have introduced a new pattern carburettor of the automatic type with single lever control. This carburettor, which is illustrated in Figs. 49 and 50, is fitted with a standard type of throttle body, but having in it a piston valve (A) to which is attached a tapered needle (B). A special jet is fitted, having a large hole bored at the top into which the tapered needle is adapted to slide.

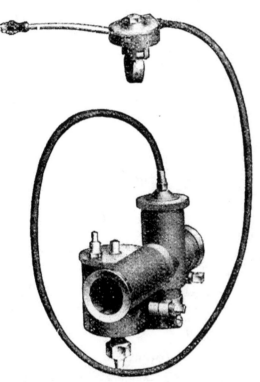

FIG. 49.—" B. & B." Automatic Carburettor (1913).

The maximum size of the jet is governed by the nozzle hole (D) drilled into the side of it ; the top portion of this jet is partly cut away and it is necessary for this cut-away portion to be facing the outlet to the engine. This can be set in any desired position by simply releasing the binding screw (E) at the bottom. The size of the jet opening is controlled by the tapered needle ; as the needle is lifted, the area becomes greater and greater, until such time as the area around the needle equals that of the nozzle hole (D) drilled in the side of the jet, when no further increase can take place. The diameter of the needle at the top of the taper is such that the jet is correct for running the engine slowly at no load. The maximum setting for the throttle in a full open position is governed by the nozzle hole (D) in the side of the jet ; all intermediate positions are obtained by making the needle of a suitable taper, so that the nozzle area increases gradually as the throttle is opened.

Behind the carburettor is fitted an extension pipe, in which is placed an automatic air valve, formed of two springs, one acting against the other. Between these springs is

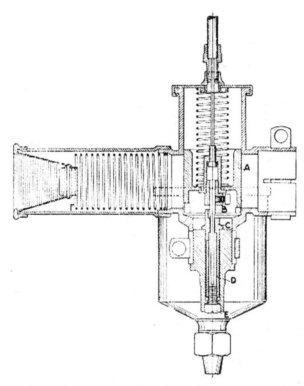

FIG. 50.—Section through " B. & B." Automatic Carburettor.

placed a small brass choke tube (F). The function of the automatic valve is to enable the engine to pick up under the load. It is quite inoperative at low speeds with the throttle almost closed, as the throttle is so proportioned that the correct mixture for slow running is obtained by proportioning the intake to the spraying chamber, the size of the jet, and the outlet from the spraying chamber.

Provision is made for slow running by a small slot on the outlet from the throttle, and the adjustment on the cabling is so arranged that one can put the lever in the shut off position and adjust the cabling until sufficient of this slot is exposed to enable the engine to turn very slowly at no load.

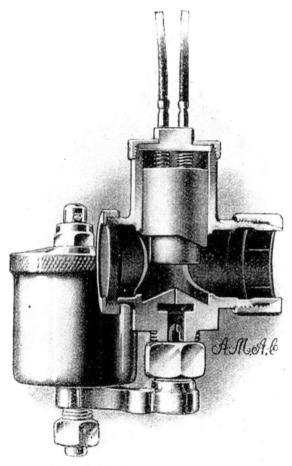

FIG. 51.—The A.M.A.C. or " Amac " Carburettor

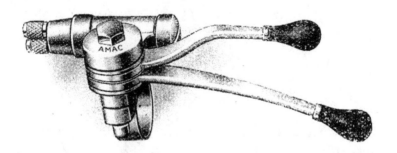

FIG. 51A.—Control Levers of " Amac " Carburettor.

This carburettor is adaptable to being set through quite a range to give various effects; for instance, it can be set to give the full power available from the engine by making the nozzle (D) equal to the biggest size that the engine can take, and then proportioning the choke tube (F) to suit this size nozzle. The tapered needle then works these two extreme points harmoniously together. On the other hand,

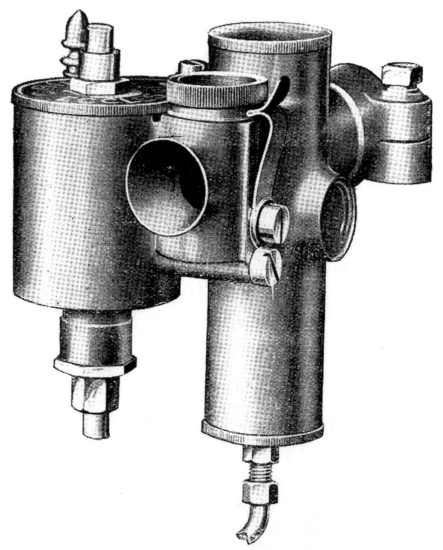

FIG. 52.—The "Binks" (Automatic) Motor Cycle Carburettor.

it can be set not to give the full available power, but to run more economically by reducing the size of the nozzle (D) and proportionately reducing at the same time the choke tube (F). This does not in any way affect the setting of the carburettor at the one end, but only limits the power available with the throttle in the full open position.

The " Amac " Carburettor.

Another well-known carburettor is the " Amac," made

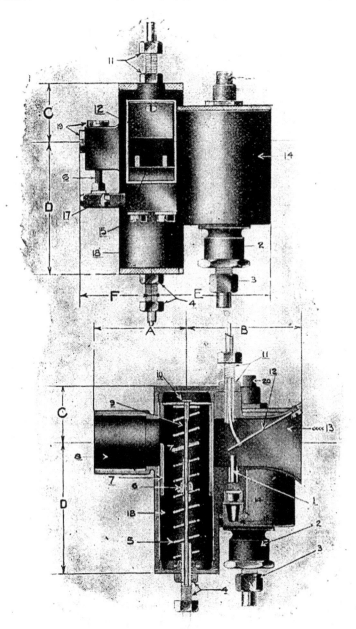

FIG. 53.—The " Binks " " Optionally Automatic " Carburettor.

by the Aston Motor Accessories Company, Ltd., of Birmingham. This carburettor is made to operate with either a single lever for both the throttle and air valves, or with a separate lever for each. The illustrations, Figs. 51 and 51A, show the double lever pattern. The mixing chamber is provided with special spraying holes varying from four to six in number, according to the horse-power of the engine for which the carburettor is adapted, and the petrol enters into a small compartment through a single nozzle stamped with its size in thousandths of an inch as a guide to the user when making adjustments in tuning up the carburettor.

This nozzle regulates the quantity of petrol let into the

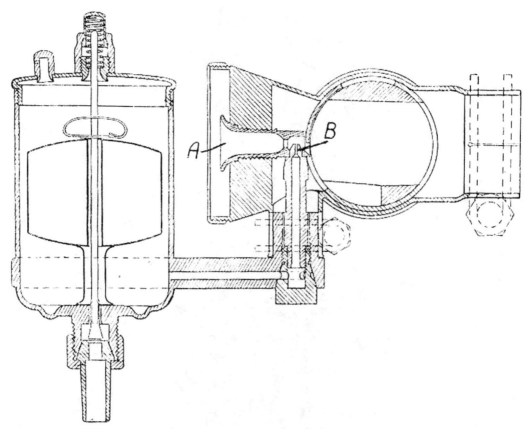

FIG. 54.—Sectional View of the "Senspray" Carburettor.

before-mentioned compartment and, as a consequence, the amount sprayed through the series of holes above it into the mixing chamber. Practically all the air is taken straight through the mixing chamber above the sprayer, and there are no bends or obstructions to interfere with the flow of the mixture. Gauzes are provided at the air intakes, these consisting of two layers of gauze separated from each other so that dust and petrol will not come in contact and thus block up the meshes.

The mixing chamber is fitted with a contracting nut to clip on the inlet. The air and throttle valves in this type are independently worked by two Bowden cables. The throttle valve shuts off the mixture very gradually. The air valve is really a variable choke tube in its action, because as the air valve closes it gradually reduces the area just above the sprayer so that the air at this point always has the correct velocity necessary for perfect atomising of the petrol. The makers claim that as the air and throttle valve to a certain extent supplement each other, this carburettor is much more automatic in its action than the ordinary two-lever carburettors, and neglecting the finer points, there are practically only three positions for the air lever—one for country roads, one for town traffic, and closed position for starting.

Since the first and second editions of this work were

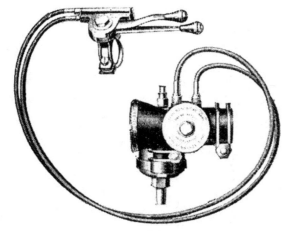

FIG. 55.—Collective View of " Senspray " Carburettor.

published, other carburettors of British manufacture have come prominently to the fore and are now being largely used on motor cycle engines. Of these the Binks, Senspray, and Lukin are entitled to special description. The first-named is made in two patterns for touring machines, one of which is constructed for wholly automatic working, and the other for either automatic or non-automatic at the option of the rider.

The principle on which the Binks automatic double jet carburettor (Fig. 52) works is that when the throttle is at the bottom the carburettor is shut off, and when it is first raised it partially uncovers the smaller or pilot jet. This gives a high velocity past the jet at low engine speeds and ensures easy starting and very slow running. The makers

state that it is possible to drive while using the pilot jet
only from dead slow to 15 miles per hour. As the throttle
is further raised the effect is to uncover the main or larger
jet. This gives more power, and as the speed of the engine
increases so does the degree of suction as the large jet, while
it decreases on the small jet, one jet thus correcting the
other ; and this, it is claimed, gives a good mixture right

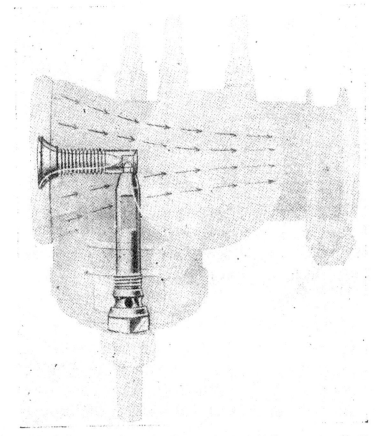

Fig. 56.—Sketch showing Action of the " Senspray " Carburettor.

up to a fully open throttle. With this arrangement it is
possible that when going at full speed the mixture may be
a shade too rich, and to correct this, if a further opening of
the throttle with the same lever is given, the effect is to
uncover the large air port by which the mixture can be
corrected.

The carburettor has the reputation of giving an extremely
rapid acceleration of the engine from standing, and also that
the throttle may be opened as wide and as quickly as
possible without choking or stopping the engine. The jets
and choke tubes are detachable in a moment so as to adapt

themselves to any condition of engine or climate, and the choke tube is capable of being varied in area by twisting it round slightly. The carburettor is controlled by a single lever, and is adaptable to any class of machine. The optionally automatic Binks carburettor, Fig. 53, gives exactly the same results, but it has an adjustable flap in the main air intake. When this flap is in mid position, the carburettor is entirely automatic. In running down hill the flap can be raised, thus practically doubling the opening into the engine; and at critical corners, when hill-climbing, it can be lowered right down on top of the jet so as to give a very rich mixture at very slow engine speeds with a wide open throttle. This pattern carburettor has two control levers, but only one need be used for all ordinary driving.

The Senspray carburettor (Figs. 54 and 55) was first introduced late in 1911. It is of the semi-automatic type and fitted with two controlling levers. The action of the carburettor is as follows :—

A small volume of air is drawn, by the downward movement of the piston, through the vaporiser or spraying nozzle (A), at an extremely high rate of speed (due to the small bore of same), over the top of the petrol jet (B), and, acting on the " injector " principle, forcibly draws the petrol out of the jet and sprays it into the mixing chamber in the form of a fine mist. At the same time the air necessary to form the explosive mixture is admitted straight in at the back of the carburettor, as illustrated diagrammatically.

The location of the jet and other parts is clearly shown in the sectional drawings, while Fig. 56 is intended to illustrate the action of the carburettor ; the path of the petrol mixture is indicated by arrows. The movements of the air and throttle valves in this carburettor is a rotary one, in place of the vertical or up and down one usually adopted.

The construction of the Lukin single-control carburettor is clearly shown in Fig. 58. In this design the air enters at the bottom of the carburettor through the gauze (P) and by the ports (A), which are controlled by the plate (B) traversing the mixing chamber (G) to the engine through the outlet (C).

The petrol flows from the float chamber along the passage (T) and up the jet spindle (R), and is sucked thence through two orifices in the jet nipple (S).

The control chamber (E) communicates with the mixing

chamber (G) through four orifices (three at F, and one at F₁) and with the atmosphere through the control port (H), regulated by the plate (I).

The plate (I) is held in position by a small keep plate (V), which has two lugs on it ; these lugs engage in the cover (O), also in the plate (I), thus fixing the two together. The setting of the plate (I) is altered at will by removing the pin at (M), rotating the cover in the required direction, and replacing the pin. The lever (L) clips on to the top of the control chamber spindle (W), and moves the whole cylinder, thus rotating the valve (B) over the ports (A) and the ports (H) under the the valve (I).

The working of the Lukin carburettor is as follows : When the throttle is closed the ports (A) are completely shut off, but the ports (H) are wide open, thus admitting air freely to the control chamber (E), and so reducing the suction on the jet (S) that only a very small amount of petrol is sucked from it through the orifices (F). As the throttle is opened to admit air through the ports (A) the whole control cylinder (as before explained) rotates, thus closing the ports (H) and admitting less air to the control chamber (E). Thus, as more air is admitted at the bottom of the carburettor, less is admitted to the control chamber, and, the suction on the jet being thus increased, more petrol is sucked therefrom and the mixture is kept correct. It will be seen, therefore, that when the throttle is wide open to admit full air supply, the control port is closed to give full suction on the jet ; when the main air port is closed for slow running, the control port is wide open to give a very small suction on the jet ; and in between these points the suction is graduated by the shape of the plate (I), so that the suction on the jet is exactly correct for every position of the main air port. No petrol passes through the orifice (F¹), the function of which is to regulate automatically the proportion of petrol and air with varying engine speed and fixed throttle position, and to control the amount of air drawn through port (H), the suction on the jet, and, consequently, the flow of petrol through the orifices (F). While the principal function of the air entering the control chamber (E) is to regulate the suction on the jet, it also accomplishes a further useful purpose. As it passes at a high velocity with the petrol through the orifices (F) it causes the liquid petrol to be broken up into an almost invisible spray, which, entering the mixing chamber (G), immediately volatilises to fill the vacuum therein.

In automatic carburettors the mixing process is rendered rather more simple by arranging that the proportion of air required to suit a given throttle opening is adjusted automatically, so that the rider, instead of having to judge, or find by means of varying the position of the extra air lever, how much supplementary air the engine will take at different throttle openings, has it done for him, there being in this case, as has been already shown, only one operating lever,

FIG. 57.—The " Lukin " Single Control Carburettor.

which controls both the throttle and the air supply. Whether the adjustments are capable of such refinement with this method as when each controlling valve is separately and independently controlled, may seem open to question ; but it is certainly the fact that " automatic " carburettors are successfully used and are identified with some of the most remarkable performances ever attributed to a motor cycle. For every-day touring the author would personally prefer the separate method of control ; but this remark must not be construed into one of prejudice against the automatic system.

Carburettor Adjustments and Control.

Regarding the broad subject of carburation in relation to the motor cycle engine, there are many points associated with the adjustment and control of the apparatus which require attention. In reality the adjustments are few, but *very* important, and, once having got them right, the old advice as to "letting well alone" cannot be too closely adhered to. The process of "tuning up" a carburettor does not, as some seem to think, follow a stated rule ; but depends very largely

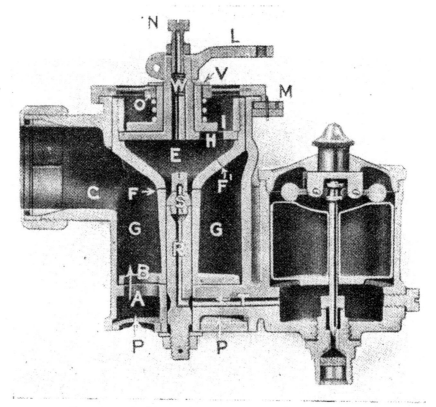

FIG. 58.—Sectional Elevation of the "Lukin" Carburettor.

upon individual circumstances connected with the type and working conditions of the particular engine ; but it must not, on the other hand, be supposed that one has to go groping about altogether in the dark as to what is wrong before being able to effect any improvement.

There *is* a process or system of adjustment based upon the symptoms displayed by the engine when working irregularly, and if these symptoms are traced beyond doubt to trouble with the carburettor, then a predetermined plan can be put into execution which must very soon elucidate the precise cause of the trouble and put it right.

When the engine shows a falling off of power, or refuses to start easily, the average rider says "carburettor" at once, and proceeds to alter the petrol level in the jet, change the jet itself for another size, or otherwise tamper with the existing adjustments ; but too much stress *cannot* be laid on the point that unless the symptoms are so pronounced, or something about the carburettor itself shows distinctly that it is indeed a fault connected with the mixing or feed of the vapour, the motor cyclist should first ascertain that it is not the charge but the means of igniting it that is wrong. In other words, let him first see that he is getting an efficient spark wherewith to explode the mixture before he does anything in the way of trying to "improve" the apparatus which produces· it ; and, further, he must also make sure that the valves of his engine are working properly, admitting and expelling the charge, before tackling the question of faulty carburation.

The explanation of the methods usually resorted to in adjusting the carburettor is reserved until the Chapter on Overhauling and Tuning up the Engine is reached, the author's purpose being that the main points connected with the adjustment and control of the engine and its principal component parts may be followed in sequence and by reading consecutive pages, and that the more detailed information bearing upon these points should be reserved until after the reader has made himself acquainted with the name and purposes of the different parts constituting every branch of the construction.

CHAPTER VII.

THE IGNITION.

THE apparatus employed for igniting the explosive mixture in a motor cycle engine is possibly less well understood by the majority of riders than any other part of the construction. They regard it as something in the nature of a mystery, and one often hears the remark, " I'm no good at electricity," made by the motor cyclist whose engine is suffering from some complaint connected with the ignition. The author well realises that the beginner is placed at a considerable disadvantage unless he possesses an average knowledge of electrical matters, and would suggest that he carefully peruses the general explanations which follow as to the methods of producing the electric spark and the appliances by means of which it is caused, to effect the purpose of igniting the charge within the cylinder.

Evolution of Ignition Methods.

Amongst the earliest forms of ignition to be used in a gas engine—which is similar in principle to a petrol engine— was the electric spark. This form of ignition has exceptional advantages ; but, at first, it was found impossible to make it reliable, and, as a consequence, the electric-spark principle gave way to flame ignition, and then to red-hot metal surface ignition, the commonest form of which is a red-hot tube, into which the gas is compressed when ignition is required. In a gas or petrol engine there is a very brief moment in which the gas should be ignited, if good results are to be obtained from the engine. Even in a slow-running gas engine it is not possible to make the ignition occur very accurately with the flame or hot-tube method, and in a petrol engine, as applied to motor cycles, the accuracy of ignition at exactly the right moment is of far greater importance, owing to the high speed at which the engine runs.

Now the electric spark can be made to occur at absolutely the correct moment, and it is chiefly for this reason that this form of ignition is practically universal on all petrol engines at the present day. Even in the early days, when any trouble occurred with such engines, it was generally safe to assume that the ignition was the cause; but the subject has received such a great amount of very skilled attention since, that the ignition apparatus is now fully as reliable as any part of the motor cycle.

Production of the Electric Spark.

There are several forms of electric ignition in use on motor cycle engines, but each system has the same essential parts, consisting as follows :—

(1) An instrument for producing an electric spark ;
(2) A prepared path for the spark to travel from the instrument producing it to the fitting on the engine where it is required ; and
(3) An appliance on the engine at which the spark occurs.

Instruments for producing the electric spark can be divided into two main classes, viz.—the coil and battery system, and the magneto system. The coil and battery system consists of a battery in which a low-pressure electric current is chemically stored, and a coil, which is an instrument for converting the low-pressure current from the battery into a high-pressure current or electric spark. There is required, in addition, a mechanical device, driven by the engine, to decide when the spark shall occur, this fitting being called a contact-maker, contact-breaker, or wipe contact, according to the various forms in which it is made. The magneto system may be said to combine the above three parts in one.

The next point to be considered is the prepared path for the spark to travel between the magneto or coil to the engine, and this is called the high-tension cable. It is convenient to think of the spark travelling down this cable, although in reality it is an electric current of small quantity, but of very high electrical pressure or tension. It has been found that an electric current will flow very easily in all metals, and only with great difficulty in such materials as rubber, porcelain, mica, or ebonite, and many others, and these last named are called insulators.

Dry air is a good insulator, and it requires a high electrical pressure to jump across even quite a small air-gap.

Compressed air is a still more powerful insulator. The high-tension cable consists of several strands of copper wire twisted together, which are completely enclosed in a thick covering of rubber.

The fitting on the engine at which the electric spark occurs is called a sparking plug, and it consists simply of a short thick wire, carefully insulated from the engine. One end of the wire is outside the engine, and is arranged with a suitable fitting for connecting it with the high-tension cable. The other end is inside the engine, and so arranged that there is a very small air-space between this wire and another one, also forming a part of the plug, and which is touching the inside metal of the engine. These two are commonly referred to as the sparking plug " points," and across the small gap by which they are separated the high-tension current has to jump. In doing so, it makes the air or gas in its path white-hot, and so produces ignition of the gaseous mixture within the cylinder.

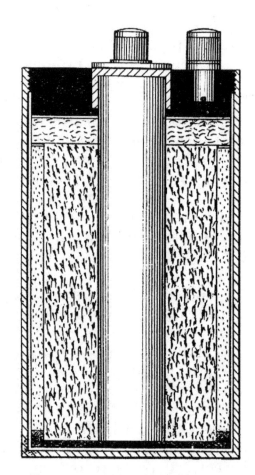

FIG. 59.—The " Hellesen "
Dry Battery.

Ignition Apparatus : Dry Battery System.

We may next turn our attention to the form and application of the different kinds of apparatus employed to convey the principle of ignition to its practical uses.

In the coil-and-battery system the battery forms a very convenient method of supplying a portable means of producing the electric spark. Batteries are divided into two main classes, called primary and secondary. Dry batteries (a form of primary cell) produce an electrical current by chemical action, and when all the current that they will give has been taken out of them they are of no further use for ignition, but may be employed for electric bells, telephones, etc., and will give long service thereon.

A primary battery generates an electric current by chemical action between the elements which constitute the battery.

The only primary battery suitable for motor ignition is the dry battery, a diagram of which is shown. It is made up with a zinc outer case, which is insulated on the outside by means of a cardboard case impregnated with insulating material, and in the centre of the zinc case is placed a carbon block, which forms the positive (+) pole of the cell, whilst the terminal attached to the outer zinc forms the negative (−) pole. Surrounding the carbon block is packed a mixture consisting of broken carbon, graphite, and manganese di-oxide ; this composition is tightly compressed round the carbon rod,

FIG. 60.—External View of the "Hellesen" Dry Battery.

and is held in position by a cloth sack, the space between this sack and the zinc case is filled in with a thick paste of jelly, which forms the active material or electrolyte ; above the sack a separate compartment is arranged, which is filled with husks, or similar material, so as to form a receptacle for any gas which may be formed during the time the cell is in action ; the upper portion of this compartment is formed by a special layer of asphalt, through which a small glass vent tube projects into the husks, so as to allow the gas to escape.

Dry batteries have the advantage of convenience, due to the fact that there is no acid to spill, no charging to look after, and no corroded terminals, and also to the fact that they are always ready for immediate use without any attention whatever.

The above is a general description of the well-known "H. H." Hellesen dry battery, for which it is claimed it does not deteriorate when not in actual use, so that the mileage can be taken from it over extended periods, thus rendering it exceptionally suitable for use on motor cars, cycles, and motor boats in general.

Accumulators and Coils.

The secondary battery does not actually produce a current, but can store it and give it out when required. Despite the fact that these batteries (or accumulators, as they are generally termed) are made with heavy lead plates, and have to contain dilute sulphuric acid, they are quite the most com-

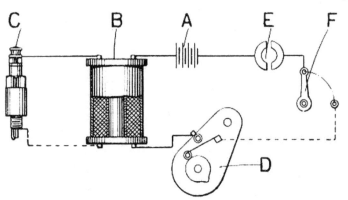

FIG. 60A.—Diagram of Wiring for Accumulator Ignition.

(A)—Accumulator. (B)—Coil. (C)—Sparking Plug. (D)—Contact-Breaker. (E)—Interrupter Plug. (F)—Switch.

mon form in use, on account of the large current that can be stored, and the fact that they can be used over and over again by re-charging, which, in other words, means simply passing an electrical current into them.

Batteries or accumulators, as arranged for motor cycle work, have two terminals, to which the high-tension wires are connected. One of these terminals is called the positive terminal, and is usually either painted red or marked with a plus (+); while the other is called the negative terminal, and is painted black or marked with a negative sign (−). It is of the greatest importance that these two terminals be never joined together by metal or wire, as this "short-circuits" the battery and does great harm.

Coils are instruments used to transform the low-pressure battery current to a high enough pressure to jump the spark-gap

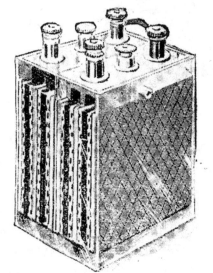

FIG. 61.—Accumulator for Motor Cycle Use.

at the sparking plug. A coil consists of a soft-iron core, round which are wound two layers of thick copper wire, outside which are a very large number of turns of very thin wire, and to produce the spark the battery current is sent round the thick-wire coil and suddenly interrupted. The coils are made up in two chief forms—*i.e.*, plain (or non-trembler) coils and trembler coils. In a non-trembler or plain coil the sudden stoppage of the battery current is performed by a mechanical device on the engine called the contact-breaker. This contact-breaker makes contact for a moment, and then suddenly breaks contact at the correct

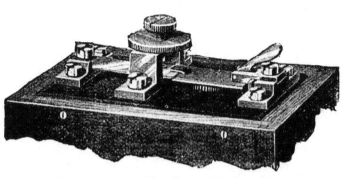

Fig. 62.—Adjusting Mechanism of Trembler Coil.

moment when the spark is required.

In a trembler coil the sudden stoppage is performed by a magnetic device on the coil itself, similar to an electric bell, and called the trembler. These coils work with a contact-maker or wipe contact, this fitting making the contact and so allowing the current to pass to the coil at the moment when the spark is required to ignite the explosive charge in the cylinder.

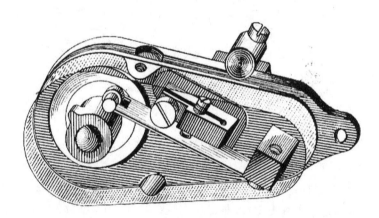

Fig. 62A.—Contact-breaker with Single Blade for Single-cylinder Engine.

Coils usually have three or four terminals, to which wires have to be connected. Two of these terminals are always low-tension and one high-tension, except when there are four terminals, and then there are two high-tension.

This number of terminals applies to single coils for single-cylinder engines. The terminals are either labelled or have

7

letters to show what part of the ignition apparatus they are to connect up with by means of the wires.

On an English coil these letters are usually :—

C = Contact Maker.
B (or A) = Battery or Accumulator.
P (or SP) = Plug or Sparking Plug.
E (or F) = Earth or Frame.

French coils usually have the following letters :—

C = Contact Maker.
P = Pile = Battery.
B = Bougie = Sparking Plug.
M = Masse = Earth.

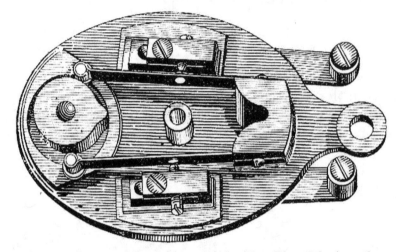

FIG. 62B.—Contact-breaker with Double Blades for Twin-cylinder Engine.

The Magneto System.

The magneto is an instrument which generates a low-tension current mechanically. This current is suddenly interrupted in a manner quite similar in principle to the ignition coil; but there is no battery, and the contact-breaker is combined with the instrument itself. The magneto system requires very little wiring, and very little attention is required to keep it in order. It is universally applied to modern motor cycles, for which purpose it is far and away the most suitable which can be devised.

The current can only be generated while the engine is in motion, and that, of course, is the only time when it is really needed. When the engine stops the magneto machine stops working also, as it derives its motion by means of a chain or gear drive from the engine itself, which rotates the armature shaft and sets the system in motion which results in producing the electric spark to ignite the explosive charge.

The diagrams (Fig. 63, 1-3), in conjunction with the remarks accompanying them, provide an easy method of studying the action of the magneto.

Fig. 1 shows the path of the magnetic lines due to the field, with the armature in the position shown, and with the low-pressure or primary winding on open circuit. It will be noticed that all the lines pass through the core of the armature, as is natural, owing to the lines much preferring an iron path to an air path. The lines, in passing through the core, also thread through every turn of wire on the armature.

Fig. 2 shows the path of the magnetic lines when the armature has turned through 90 degs. All the magnetic lines now

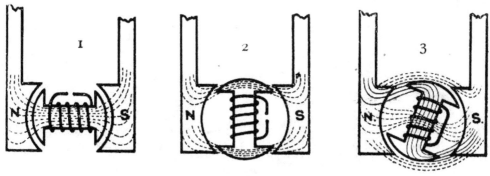

(N)—North Pole. (S)—South Pole.

FIG. 63.—Showing the Action of the Magneto and Path of the Magnetic Lines.

pass through the ends of the armature, but to get to this position they have had to cut every turn in the armature winding. If the winding had been a closed circuit, a current would have been produced in the winding, the amount of the current being dependent on the resistance of the turns and the pressure developed. The pressure produced depends on the strength of the magnetic field, the number of turns in armature primary winding, and the rate at which the magnetic lines are cut by the winding, or, in other words, the speed of the armature.

When the armature has revolved through another 90 degs., the path of the magnetic lines is as shown in Fig. 1 again; but, as the field is always in one direction, it will be observed that the lines pass through the winding in the reverse direction. For this reason, when two sparks are produced in each revolution, the sparks will be in alternate directions.

It should be clearly understood that the above diagrams only show the path of the magnetic lines on an open armature circuit. If the winding is closed and the armature revolving, the current is produced in the winding, the greatest current being when the armature is in *about* the position shown in Fig. 2. But this current round the armature core produces a magnetic field of its own. This armature magnetic field distorts the permanent magnetic field in the direction of rotation, and makes the point of strongest armature current, and therefore strongest armature field, later than shown in Fig. 2.

It is this distortion of the magnetic field that makes the design of the advance and retard of magneto ignition so awkward. Much the simplest and neatest form of timing the

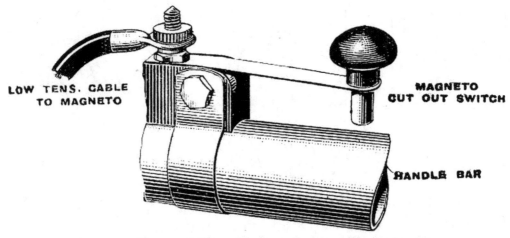

LOW TENS. CABLE
TO MAGNETO

MAGNETO
CUT OUT SWITCH

HANDLE BAR

FIG. 64.—Another Type of Handle-bar Cut-out Switch ; Magneto
Ignition.

ignition is to revolve the contact-breaking points through a small angle, so that they either separate a little earlier or later than usual. As the position where it is suitable to break the current extends over a very small angle, and as the distortion of the field occurs in the same direction as the rotation, while to advance the ignition the contact points have to be revolved against the direction of rotation, it will be appreciated that the range of movement is still further limited. The result in practice is that, for starting, the break has to occur in rather a bad position ; and, owing to the speed of the armature being slow at starting, it explains that it is sometimes necessary to advance the spark, even for starting, in order to get a good enough firing spark. It is for the above reason that many machines have the timing of the ignition fixed.

Fig. 3 shows the magnetic field with the armature primary winding closed, and revolving at a good speed, the armature being in the position just before the contact points are going to separate. The lines of force of the permanent magnetic field are shown by dotted lines, and the lines due to the armature current are shown in full.

It will be seen that the armature field opposes the permanent magnet field, and this explains the de-magnetising

FIG. 65.—The "Lodge" Magneto Cut-out affixed to Handle-bar.

effect that occurs if the armature is short-circuited by means of a switch, so as to use the engine as a brake. There is probably no harm in switching off the magneto in this way, but the magneto should not be driven at a high speed, as when coasting down a long hill with the armature short-circuited. If, however, the makers of magnetos arranged so that the switch left the armature circuit open, there would be no harm in using the engine as a brake. As it is, the Bosch Magneto Company claims there is no risk and that the life of the magneto is not appreciably decreased by the use of the switch, or, as it is commonly called, the magneto "cut-out," an illustration of which, attached to the handlebar of a motor bicycle, is given in Fig. 64. Another type is shown in the next illustration, Fig. 65. This is a neat "concealed" pattern switch for attachment to the handlebar, and has the advantage of being quite weather-proof. By pressing down the black button on top of the switch, the ignition is cut out and the engine stops working. The fitting is an all-metal nickel-plated one, and is made by Messrs. Lodge Bros., of Birmingham.

From an examination of this diagram, the quick reversion in the armature field that takes place as soon as the contact

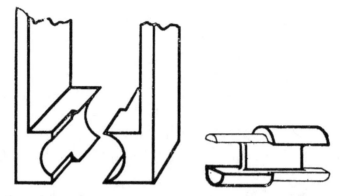

FIG. 66.—Armature and Pole-pieces for V Type Twin-cylinder Engines.

points are separated will easily be understood. As soon as the contact points separate, the current in the armature circuit is stopped, and there is nothing to sustain the armature field, and therefore the permanent magnetic field, which it has been opposing, at once overpowers it, thus causing ex-

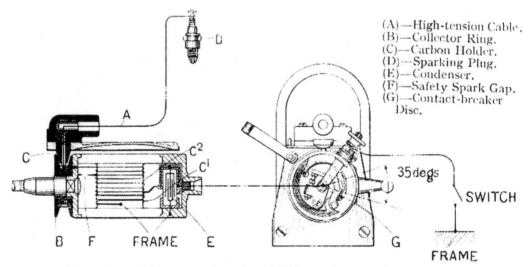

(A)—High-tension Cable.
(B)—Collector Ring.
(C)—Carbon Holder.
(D)—Sparking Plug.
(E)—Condenser.
(F)—Safety Spark Gap.
(G)—Contact-breaker Disc.

FIG. 67.—Diagram showing Wiring Connections, Bosch High-tension Magneto Ignition.

ceedingly quick reversal of the magnetism, and inducing a very powerful spark.

The power taken to drive a single-cylinder magneto is about 1-15th h.-p. at normal maximum speeds. The power required increases roughly as the square of the speed ; but not exactly, owing to distortion of the field.

In a single-cylinder magneto there are two maximum positions ; but only one position is used, as the armature is kept on open circuit by the platinum points being kept apart as the armature revolves through this position.

In the case of two-cylinder V engines the correct timing is obtained by making specially shaped pole-pieces and a specially shaped armature, as indicated in Fig. 66. The effect of this arrangement is to make the two maximum positions occur at the correct angle for firing each cylinder, instead of being exactly opposite, as they would be with a normal magneto.

Construction and Working of the Bosch Magneto.

Passing on from this to a consideration of the detailed constructions of the magneto and the functions of the various parts the diagram (Fig. 67)—for the use of which the author is indebted to the Bosch Magneto Company, Ltd.—clearly shows the wiring and points connected with the magneto system of ignition, and the details of the apparatus can be followed by referring to the other illustrations.

The principle of operation is as follows : Between the pole shoes of two steel magnets which form a strong magnetic field, a so-called shuttle armature rotates, and by reason of this motion a current is produced in the armature winding, which reaches its maximum twice in one revolution, but one of which is used only for effecting a spark between the electrodes or points of the sparking plug. The jumping of the spark across this air-gap has already been explained. The armature is built up of thin sheets of metal securely held together, and is wound in two parts—one forming a continuation of the other, and of which the inner winding (C^1) is the primary, consisting of a few turns of heavy wire ; and the other, the secondary (C^2), consisting of many turns of fine wire. The electrical current is produced in the primary winding by the rotation of the armature, and, by the interruption of the primary circuit by means of a contact-breaker, a high-tension current is induced in the secondary winding, which at the moment of the separating or breaking, as it is termed, of the contact point causes a spark to jump across the electrodes of the sparking plug and fires the charge within the cylinder of the engine.

As the arc-like spark can only be produced when the armature is in a certain position, and as the ignition has to take place at a certain period in the movement of the piston, it becomes necessary that the armature shall be positively driven at the same speed as the cam shaft of the engine. To effect this a chain or gear-wheel is affixed to the end of the cam shaft of the engine, and another chain or gear-wheel—of the same diameter and pitch exactly—to the end of the magneto armature shaft, and the two are connected

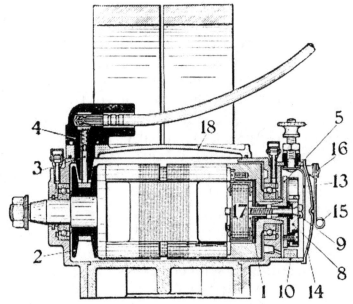

Fig. 68.—Section through Armature and Contact-breaker of Bosch Single-cylinder Magneto.

up, either by a chain or intermediate gear-wheels, as the case may be, so that the armature shaft is bound to rotate at the same speed as the cam shaft.

This gearing is shortly referred to as the " magneto drive," and is enclosed within an aluminium casing to protect it from dust and dirt.

The variation of the moment at which the spark shall occur within the cylinder, in accordance with the position of the piston working therein, is effected on the magneto itself by means of the timing lever, causing the interruption of the primary current to take place earlier or later in relation to the stroke of the piston.

This is called the timing of the spark, and it exercises

a very considerable influence on the working of the engine. A lever or small handle—termed the firing or spark lever—is affixed to the tank or handle-bar of the motor bicycle, and this connects with the contact-breaker by means of a rod or Bowden wire attached to a short arm on the steel segment, which influences the making and breaking of the contact points. The precise manner in which the various parts work, and the part each plays in producing the electric spark and controlling the periods of its occurrence, may be best described as follows. Referring to the drawings, Figs. 68 and 69, in conjunction with one another:—

The contact-breaker disc (5) rotates with the armature, and is directly coupled to it; but is detachable, and held

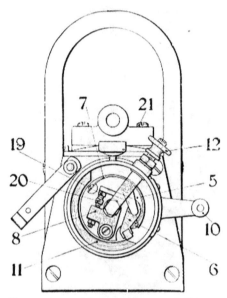

FIG. 69.—End View of Bosch Magneto with Contact-breaker Cover Removed.

by means of the screw (9) into the rear end of the armature spindle, which is bored to receive it. The contact-breaker disc (5) is prevented from rotating in the bored end of the spindle by means of a key, and keyway, which also make it impossible to place the contact-breaker in a wrong position. On the disc (5) the bell-crank lever (6) is fitted, carrying at one end a small platinum point, attached by means of a nut and screwed portion, and at the other extremity a fibre segment projecting laterally. The platinum screw on the bell-crank lever aforesaid is pressed against a similar platinum screw on the contact-piece (8) by means of a spring (7). The contact-piece (8) is fastened to the contact-breaker disc (5), but is insulated therefrom, and is electrically connected by the screw (9) to the contact-plate (1), and consequently with the end of the primary winding.

The screw (9), which holds the complete contact-breaker in position, enables it to be readily removed for the purpose of cleaning the platinum points, and this method of construction facilitates the removal of worn or damaged parts and the replacement of the same.

We now come to the timing lever, to which reference has
already been made. The timing lever (10) is held on to the
end plate of the magneto by means of a shoulder turned
thereon, so that it can rotate through
a certain distance, to effect an advance and retard in the moment of
firing. Attached to the timing lever
(10) is the steel segment (11), and
inside this slip-ring the contact-
breaker revolves. The parts (10) and
(11)—that is, the timing lever and
steel cam—form one piece with a

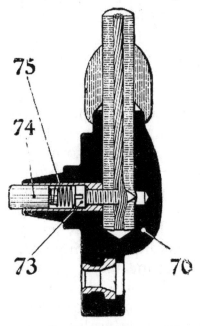

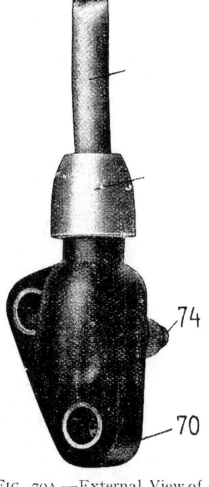

FIG. 70—Sectional View of Bosch
Carbon Holder, and with High
Brush Tension Wire Fixing.

FIG. 70A.—External View of
Bosch Carbon Holder and
High Tension Cable.

brass ring, and the whole can be rotated, within certain limits,
around the contact-breaker, whether the latter is stationary
or in motion, so altering the moment in the process of rotation
at which the bell-crank lever comes in contact with the steel
segment, and thus causing the spark to take place earlier or
later in the stroke of the piston as may be desired.

The bell-crank lever (6) is connected through the contact-
breaker disc (5) with the armature spindle, the armature core,
and one end of the primary winding; whilst the contact-piece
(8) is connected with its other end, and when the fibre head

of the bell-crank lever reaches the steel segment, the lever (6) is deflected, and the primary circuit broken by separation of the two platinum points. If, when the interruption of the circuit occurs, the armature is in a suitable position, the voltage of the current induced in the secondary winding will be sufficiently high to cause a very intense spark to pass between the points of the sparking plug.

The Condenser.

Other parts of the construction shown in the drawings,

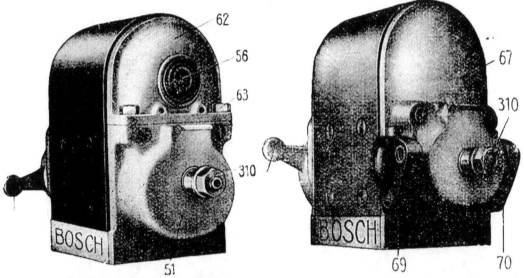

FIG. 71.—Bosch "Watertight" Magneto for Single-cylinder Engine.

FIG. 72.—A similar Magneto for Twin-cylinder Engine.

Figs. 68 and 69, include the condenser (17), the function of which is to prevent the arc due to the current induced in the winding by the collapse of the armature field. The condenser is connected across the two contact-breaking points and consists of a number of thin metal plates separated by paraffin paper or mica. Every alternate metal plate is electrically connected to another, and the set of alternate plates to one platinum point. The other set of plates are electrically connected to the remaining platinum point.

The condenser, after charging, discharges itself into the low-tension winding, causing the magnetic field to collapse exceedingly quickly. The current from the discharge of the condenser being in the reverse direction, a magnetic field is developed in opposition to the previous field.

Collection and Distribution of the Current.

The slip ring with distributing segment (2) provides a means of distributing the high-tension current generated by the magneto to a collector or carbon holder (4), this part consisting of a small brass tube with a solid portion at the top, above this being a brass knob on to which one end of the high-tension cable is " snapped " in a manner rather similar to a glove fastener. The little brass tube contains a small coil spring into which is inserted a rod or stick of carbon. This rod of carbon (or carbon brush, as it is called) is in contact with the distributor segment (2), being kept up to its work by the little afore-mentioned spring. The electrical current thus passes through the carbon holder to the cable, and by this

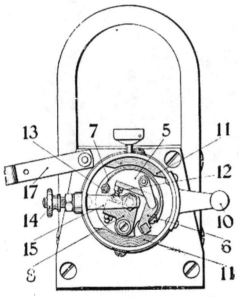

FIG. 73.—Contact-breaker with Double Segments for Twin-cylinder Engine.

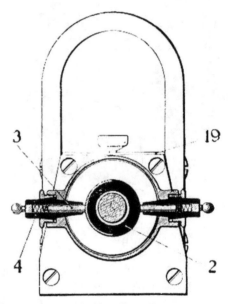

FIG. 74.—Collector Ring and Carbon Brushes for Twin-cylinder Magneto.

means to the sparking plug. The latest Bosch magnetos are so designed as to be impervious to weather conditions, and the accumulations of wet and grit upon them has no effect on their working. This type is known as the watertight magneto, and, as the illustrations show, the horseshoe magnets are now boxed in at the sides with aluminium covering plates instead of being left open as heretofore. A different method of connecting the high-tension cables has also been adopted. The carbon holder takes the form of a vulcanite block (Figs. 70 and 70A) attached to the magneto by two countersunk tapered screws, in place of the screwed portion on the brush holder itself as was formerly the case. To remove the cable or clean the carbon brush it is now necessary first to remove these screws and, by so doing, detach the whole block from the magneto. Then, to get the cable itself adrift, one has to withdraw the carbon brush (74) with its spring (75) when, at the end of the cavity a tiny brass screw (73) will be observed, and this, in turn, must be withdrawn, after which the cable can be immediately pulled out. The diminutive brass screw cuts right through the cable and its wire centre, and forms a means of communication for the current to pass to the sparking plug. It is essential that the cable should be smeared with paraffin wax before being pushed into the carbon block, so as to form a waterproof and damp-tight seal, otherwise drops of water or moisture in one form or another may reach the contact portions and result in leakage of current and short-circuiting.

With this type of magneto on a twin-cylinder engine it is more often than not a very difficult matter to get at the holding screws at the rear of the magneto owing to the position of the latter. In nine cases out of ten some obstruction exists, and although a bent screwdriver-blade will sometimes facilitate the task, it is nearly always necessary to dismount the magneto from its platform—with resultant upsetting of the spark timing—when for any reason the carbon block or brush needs examination. This, however, is a very infrequent occurrence, but should the necessity arise by the roadside, and especially at night, the motor cyclist has anything but an enviable task before him. The screws are not set vertically one above the other, but staggered, and one of them is as a rule practically inaccessible altogether with the magneto *in situ*. So far as the generation of current is concerned, it is immaterial which way the armature revolves, this point depending upon circumstances associated with the precise method of driving the magneto from the engine. The screw (12) (Fig. 69) is provided so that a length of low-tension insulated wiring may be led from the magneto to a switch on the handle-bar of the machine, so that the current may, if desired, be diverted from the sparking plug while the magneto is working. This forms an additional means of controlling and cooling the engine. The contact-breaker mechanism is protected by a dust-proof cover (14) held in place by the spring (15), which latter is turned sideways when it is desired to remove the cover for the purpose of inspecting the contact-breaker parts. Oil wells are provided on each side of the magneto to permit of the armature bearings being lubricated, this being only required at the rate of a few drops once a month with a ball-bearing magneto. By this time the reader will probably have gained a fair idea as to the manner in which the electric spark is produced, and by what means it is conveyed to the interior of the cylinder for igniting the charge, and will be wanting to know something of the precise method in which variation in the moment of sparking is effected and the result which such variation has on the working of the engine.

The Timing of the Spark while Riding.

As before said, there is provided for the rider's use on the tank of the motor cycle a small handle, attached to which is a rod, which, at its lower end, connects with the arm or timing lever (6) on the slip ring, which carries the steel segment (11), and which may be moved within certain limits around the contact-breaker. It has also been explained that the movement of the slip ring alters the relative positions of the steel segment and the fibre block on the end of the bell-crank lever, so that the two come in contact with one another either earlier or later in the path described by the contact-breaker points in rotating. As the coming in contact of these two parts effects the separation of the platinum points (thus interrupting the circuit and causing the spark to take place), it will be seen that as the moment of separation is varied so must the moment of sparking be also varied ; and, as the armature of the magneto is caused to rotate at the same speed as the engine camshaft, it naturally follows that the piston in the cylinder will have assumed a different position by the time any fresh adjustment of the timing lever has been made, so that the explosion occurs at an earlier or a later period in the piston stroke, according to whether the timing is advanced or retarded.

The variation of the movement is, of course, somewhat limited, matters being so devised that timing lever (10) carrying the steel segment (11) can be rotated through an angle not exceeding 35 degs., which is equal to a variation of 70 degs. on the engine.

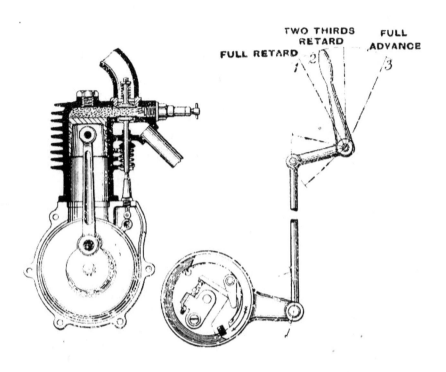

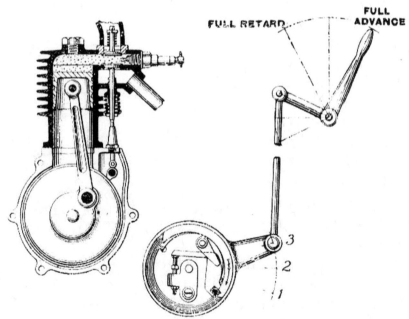

FIG. 75.—Method of Timing the Ignition
(Magneto System).

Now, as to the purpose and effect of varying the moment of sparking. It might be supposed that the precise and only moment at which the mixture or gaseous charge ought to be exploded within the cylinder would be when the piston arrives at the end of its upward (compression) stroke and when the point of highest compression has thus been reached, so that as powerful an explosion as possible might result. If this condition were to be regarded as correct, and desirable under every condition of working, then there would be no need to provide any means of varying the sparking moment ; and, indeed, a few makers *have* arranged their engines with fixed ignition and the rider has no means of altering it at his disposal. This method, like automatic carburation, may be desirable with some engines, but it is certainly not in the majority ; and, speaking personally, the author much prefers to have both the ignition and carburetting arrangements designed for variation at will.

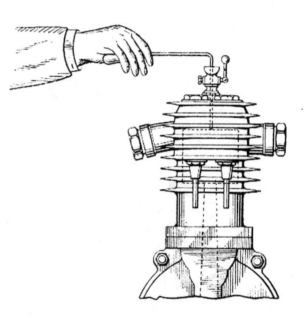

FIG. 76.—A piece of Wire, Bent and Passed through the Compression Tap, locates the position of the Piston.

Let us assume that the piston having drawn in the charge of mixture is moving upwards on the compression stroke. The gases are most highly compressed when the piston reaches the highest point of its travel, but it is not desirable to leave the moment of sparking until then, for there must be an appreciable, though very minute, lapse of time between the moment at which the platinum points separate at the contact-breaker, the spark is produced at the sparking plug points, and the explosion of the gases actually takes place ; so that if the spark is timed to occur slightly *before* the piston reaches the end of its upward stroke a complete combustion of the charge is effected and the force of the explosion is exerted at the most favourable moment for driving the engine at high speed.

So far we have only considered the timing of the spark as effected by the rider within the limits permitted by the tank control handle. There is another and even more important aspect of the matter.

The Permanent Setting of the Ignition.

As explained, every motor cycle engine fitted with magneto ignition is arranged so that the armature, and with it the contact-breaker, rotate in a certain co-relation to the movements of the piston, and this is effected by adjusting the chain or gear-wheels which serve to convey motion from the engine to the armature. The platinum points of the contact-breaker are set to break contact with one another just as the piston reaches the top of the compression stroke ; this setting, however, being subject to the rider's manipulation of the handle, as before explained. It is an easy matter to adjust the timing of a magneto. First of all, the thing is to place the piston at the top of the compression stroke— that is to say, when it returns to the top of the cylinder and *both* valves are closed, keep it in that position ; then rotate the contact-breaker by hand until the points are just on the point of separating, and with these two positions assured the gear or chain drive can be connected up. Before attempting to do anything with the other parts, however, place the firing handle on the tank or handlebar at two-thirds retard position (Fig. 75) (with some engines it is possible to nearly fully retard the handle), and, of course, the more retard given at this juncture the earlier will the spark take place in the compression stroke when the handle is fully advanced, because of the wider range of advance movement afforded. If the spark is caused to take place too early in the compression stroke—that is to say, while the piston is too far off the end of its upward travel—the effect is to produce an explosion some time before the motion of the piston can be reversed, so that we have the piston being forced upward by the fly-wheels and connecting-rod, while the explosion at the same time is trying to force it downwards ; it is, in other words, caught between two opposing forces, and this, as before pointed out, leads to what is termed " knocking " in the engine bearings, due to the jarring effect of what we may perhaps best call a collision between the opposing forces in the cylinder, and which jarring is conveyed downwards by means of the

8

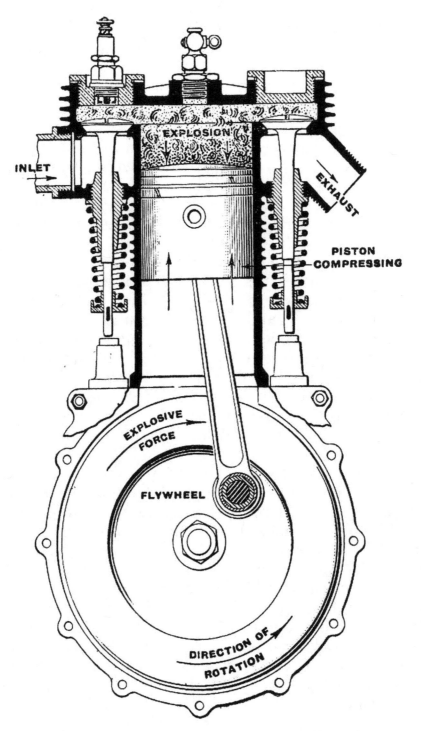

FIG. 77.—The Meaning of a " Backfire."

connecting-rod. This knocking occurs at low engine speeds with spark lever much advanced ; and is undesirable enough, but an even worse evil, *i.e.*, what is termed " backfiring," may occur at *starting* when the ignition is much too far advanced. In the event of a severe back-fire taking place it is quite possible that serious damage may result to the connecting-rod, timing gear, or other parts of the mechanism. The effect of the premature explosion is to suddenly and forcibly reverse the direction in which the flywheels are rotating, and the piston is subjected to a violent blow while still travelling upwards. Pre-ignition, or firing of the charge before the spark is timed to occur, may be brought about by the presence of heavy carbonised deposits, which under certain circumstances may assume an incandescent state of heat, or a small projection or irregularity in the cylinder casting may in like circumstances effect the same result. The accompanying diagram, Fig. 77, shows clearly what takes place when a back-fire occurs.

If, on the other hand, the timing of the spark is too far delayed, or retarded, the explosion occurs when the piston has performed a material portion of the downward working stroke, so that before complete combustion is effected and the full force of the explosion is spent, the exhaust valve opens, the piston rises on the exhaust stroke, and the gases are expelled while still containing a lot of driving power. Loss of power and overheating of the engine follow, and the engine speed falls below what it should be. It is, therefore, necessary to manipulate the firing lever so that it is fully advanced, or very nearly so, for fast running, when the machine is well under way, and can be allowed to go along at a good rate of speed, and to slightly retard the timing for starting or when otherwise it is either obligatory or necessary to materially reduce the speed of the engine for some time.

So that what it amounts to is this : At starting, and when it is desired to run slowly, slightly retard the timing lever ; and when speed has been gathered, fully advance the spark and keep it there, so that full advantage may be taken of the force of explosion in driving the engine, only altering the position of the timing lever to one of more or less retard when circumstances call for it. Chapter XII, containing some hints on driving, will further emphasise these points.

When setting the ignition with a magneto, either one of
two plans may be employed. In the case of a chain drive,
the chain can be removed altogether and coupled up in place,
after the sprocket on the engine and that on the armature
shaft have been got into the right positions relatively to one
another, or the sprocket on the armature shaft may be
loosened without interfering with the chain itself, and the
shaft revolved to get the right position; the chain wheel
being then locked on the shaft.

In a gear drive the intermediate gear-wheel (or wheels)

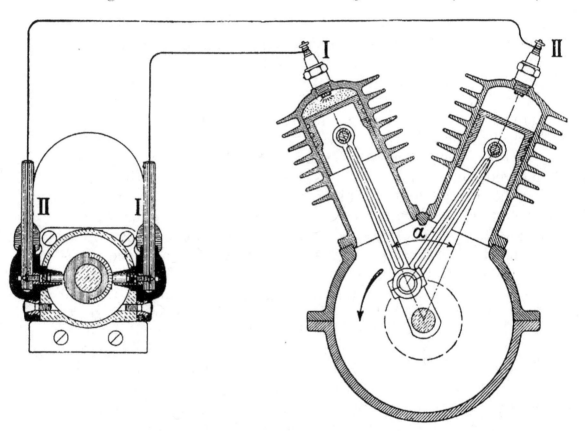

FIG. 78.—Wiring and Magneto Timing for Twin-cylinder Engine.

may be taken out, and then, when the correct positions of
the engine and magneto parts have been obtained, these
gears replaced; or, as before, the gear-wheel on end of
armature shaft can be slackened off and afterwards tightened
up again when all is correctly positioned. This last-named
method permits of more minute and accurate adjustments
being made as it is not dependent in any way upon the
" pitch " of the teeth.

In the case of a twin-cylinder engine, the contact-breaker is fitted with two steel segments, one for each of the two cylinders, so that as the contact-breaker rotates the bell-crank lever twice comes in contact with a steel segment during every revolution, and the platinum points consequently break contact twice, at set (but not regular) intervals, causing an electric spark to occur in each cylinder in turn every time the contact-breaker performs a complete revolution. In timing the ignition of a twin engine, it is only necessary to employ the same means as in a single-cylinder one, the rear cylinder being usually selected for the purpose. Then when all is coupled up it will be found that if the timing is correct as set for the rear cylinder it is correct for the front one as well.

In order that the part of our subject now under consideration may be the better understood, it may be well at this point to make a few general remarks upon

The Theory of Ignition.

It is well to get a clear idea of what actually goes on in the cylinder after compression. The usual plan is to ignite the gas at one point, and from this point combustion spreads comparatively slowly at first, with increasing rapidity as the unburnt gas becomes heated up and more highly compressed, and also as the surface of the igniting wave increases. The early stages of ignition are slow, because the igniting surface is exceedingly small; the initial stages are quicker with a large spark, which has a large igniting surface, and quicker still when two simultaneous sparks are used. The point at which ignition is started is also of considerable importance. This position should be chosen so that a good mixture is ensured in the neighbourhood of the sparking points, and the gas should be ignited as far away as possible from cold metal surfaces. With the above conditions fulfilled, and when the gas is ignited at one point, certainly the best position is in the centre of the compressed gas. Of course, a gas might be chosen, and such igniting arrangements made, that the expansion of the gas would be too sudden for the present speed of petrol engines; but at present it is rather the other way, and, as has already been stated, in order to get the best result, the spark has to be timed to occur a

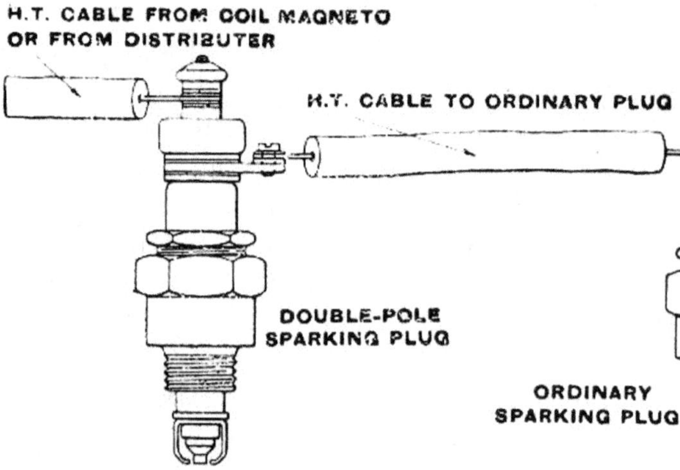

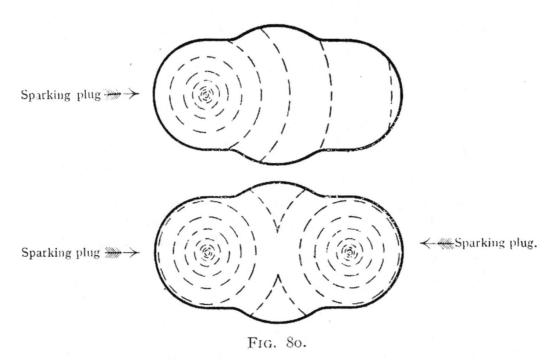

H.T. CABLE FROM COIL MAGNETO
OR FROM DISTRIBUTER

H.T. CABLE TO ORDINARY PLUG

DOUBLE-POLE
SPARKING PLUG

ORDINARY
SPARKING PLUG

FIG. 79.—The " Lodge " System of Double-pole Ignition.

Sparking plug

Sparking plug

Sparking plug.

FIG. 80.

EXPANSION OF GASES DIAGRAM.

Upper View : Charge Ignited by a Single Plug. Lower View :
Charge Ignited by " Lodge " Double-pole System.

shade before maximum compression. If the ignition arrangements were better, the charge might be ignited exactly at full compression, or, at any rate, not so much before full compression, with a consequent increase in power. There is an undoubted advantage in igniting at different points, for then the combustion has a smaller distance to travel from each igniting point, and the early stages of combustion are accelerated, allowing the timing of the spark to be made a shade later Fig. 80 illustrates these remarks.

Double-Pole Ignition.

The author, with these thoughts in mind, made a trial some time ago of the " Lodge " system of double-pole ignition on a motor cycle engine, and experienced a marked improvement in the running of the engine so fitted. Starting from cold was materially assisted, and owing, no doubt, to the more perfect and rapid combustion of the charge, it was possible to run with less throttle opening than before, thus economising petrol and assisting to keep the engine cool. The system is a very simple one, consisting of a double-pole sparking plug, which is screwed into the cylinder in the ordinary way, and a second plug, of standard construction, placed over the exhaust valve or where otherwise convenient. In the engine used the double-pole plug was placed over the inlet valve, and the ordinary plug over the exhaust valve, and the high-tension cable led from the magneto to the double-pole plug. Then a short length of high-tension cable was used to connect up the two plugs, a special fitting being provided on the double-pole plug to allow of this.

A diagram of the arrangement is shown in Fig. 79. With this system in use, the mixture within the cylinder is ignited at two separate points instead of only one, so that more rapid combustion takes place. Increasing the rate of combustion enables the maximum pressure on the piston to be reached earlier in the working stroke without the back-pressure and overheating that would result if the same effect were attempted with one plug, by setting the time of firing so much earlier. Obtaining the maximum pressure early in the stroke gives increase of power, the effect being most noticeable at high speeds.

It may be suggested that much the same result could be arrived at by connecting up two ordinary sparking plugs in parallel, and thus dividing the current from the coil or magneto terminal into two circuits. It becomes a question as to whether a sufficiently intense spark can be obtained at both plugs, and also whether the two will synchronise in the distribution of the current, for, of course, if they fail to do so, the force of the explosion will, as a consequence, suffer; and we may take it that even if a sufficiently intense spark *could* be obtained in these circumstances it would more likely than not be useless, owing to the impossibility of adjusting the gaps of the two plugs to exactly the same distance. The plugs must be in series to obtain synchronism. The mixture must be exploded in the combustion chamber at one precise moment to gain maximum power, and not in successive stages, however finely divided.

Ignition Systems Compared.

In a very few cases motor cycles are fitted with low-tension magnetos, and, where this is done, it is necessary to employ a coil, so that, as in the case of an accumulator, the low-pressure current may be converted into a high-pressure one, this being absolutely essential to the production of a sufficiently intense electric spark. The vast majority of motor cycles are fitted with high-tension magnetos, with double-wound armatures, so that no coil is needed, and there can be no doubt whatever that the latter system is in every way the better one, both on the score of simplicity and efficiency in working.

Each system of ignition has some advantages, and, of course, some disadvantages; but the latter, in the case of the high-tension magneto, are wholly outweighed by the advantages offered.

To sum up the relative claims made on behalf of each system, it may be said that the non-trembler coil requires very little battery current, so the batteries last a long time. It gives a very accurate timing of the ignition, and consumes no power from the engine. Care must be taken that the engine is not stopped in such a position that the current is left on, as if the switch is closed when the engine is standing in this position, the batteries are sure to be damaged.

The trembler coil affords much the best system for starting and slow running; but, for very high-speed work, the timing of the ignition is not quite as accurate as the other systems.

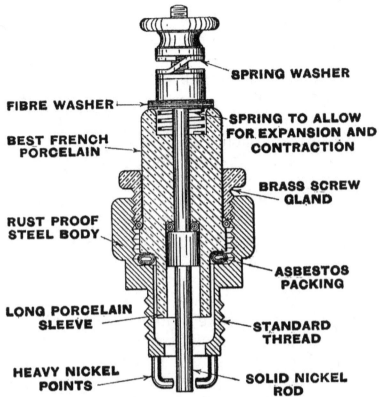

SPRING WASHER

FIBRE WASHER

SPRING TO ALLOW FOR EXPANSION AND CONTRACTION

BEST FRENCH PORCELAIN

BRASS SCREW GLAND

RUST PROOF STEEL BODY

ASBESTOS PACKING

LONG PORCELAIN SLEEVE

STANDARD THREAD

HEAVY NICKEL POINTS

SOLID NICKEL ROD

Fig. 81.—Sectional View of "Lodge" Sparking Plug.

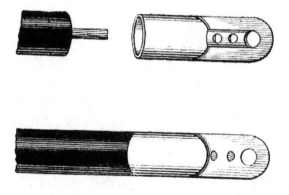

Fig. 82.—High-tension Cable Wire and Terminal shown Connected and Disconnected.

It also consumes more current than a plain or non-trembler coil.

The high-tension magneto is by far the neatest, most compact, and simple to operate of any system. With it no batteries are required, and the wiring is of the most simple character, consisting in fact of only one high-tension insulated wire from the collector terminal on the magneto to the sparking plug. Practically the only disadvantage about a magneto is that it takes a certain proportion of engine power to drive it ; but this is small enough to be considered negligible.

Care of the Ignition Appliances.

Longer and more satisfactory service may be obtained from the various appliances employed for ignition if care be taken to maintain the same in clean and proper working condition, and, with a view to assisting the reader in this direction, the author has prepared the following notes, covering the full range of equipment previously described :—

Sparking Plugs.— As before said, there are a large number of different types of sparking plugs, these plugs being insulated with either porcelain, steatite, or mica. The plan is sometimes recommended of using very cheap sparking plugs, as a number of these are supposed to be more economical than one good one ; but this is a bad plan, as, even if it should be more economical (which is doubtful), the plug may fail at a most awkward moment. Much the best plan is to buy plugs of a well-known make, and if the author's long experience stands for anything, the utmost reliance may be placed upon the Bosch, Lodge, and Sphinx ignition plugs. Noteworthy among these is the self-cooling plug of Messrs. Lodge Bros. & Co., which fully justifies the claims made for it.

The Sphinx Manufacturing Co., of Birmingham, supply a range of sparking plugs for motor cycle use, any one of which may be safely relied upon for good and long-continued service. The " Gnat," one of this series, is a specially shallow plug, *i.e.,* one-inch high above cylinder, and thus adaptable to any position that may in special cases be desired. The other plugs are of varying heights, affording different lengths of reach, and all are well made and finished in the best style. The author has used these plugs for a considerable time past, and has found them excellent in every way.

If through faulty lubrication of carburation the plug gets very dirty on the side exposed to the flame, the best

Fig. 86.—Another Pattern of "Sphinx" Sparking Plug.

Fig. 85.—"Sphinx" Sparking Plug.

Fig. 84.—"Lodge" Self-cooling Sparking Plug with Ribbed Exterior.

Fig. 83.—"Bosch" Sparking Plug.

way to clean it is to pour petrol into the plug and clean it inside with a small stiff brush. It is of great importance that the plug should be quite gas-tight, and the best way to test for this is by smearing round the joints with oil and then pedalling the engine round, when, if compression is leaking, small bubbles will appear where the oil is smeared on. The plug should be screwed down on to an asbestos washer. With frequent screwing up these copper asbestos washers get hard, and do not make a proper gas-tight seat. They can then be annealed by heating them up to a dull red heat and allowing to cool.

The best distance apart of the sparking points is 0·5 mm. Closer than this would be better for starting; but with a powerful magneto at high speeds, if the points are set any closer there is danger of them fusing together.

A sparking plug should be chosen with a length of reach into the cylinder such that the spark occurs about flush with the inside surface. If the plug projects too far into the cylinder it is liable to get very hot; and if the spark occurs at the back of an opening because the plug has too short a reach, misfiring is very liable to occur, and possibly pre-ignition.

Wipe Contact (for use with trembler coils).—It is important that this should be kept clean, and that there should be a good spring press contact. A little vaseline or oil can be used for lubrication. There should be no looseness in any of the parts, and the surface of the wipe must be perfectly true. The contact-pieces should extend for about 40 degs. In many wipe contacts the " earth " return is through a shaft. This is liable to produce bad contact, and a proper " earthing " terminal is a distinct improvement.

Contact-makers (for use with trembler coils).—Much the best design is that in which the contact is closed positively, and separated by means of a spring. The contact points are usually of platinum iridium, and if the current can be reversed in direction regularly through these contacts, it will prevent them pitting. All the parts should be kept thoroughly clean, and only the cam actuating the contact points will require a little lubrication.

Contact-Breakers (for use with non-trembler coils).—It is essential that the contact points be made of platinum, or, better, platinum iridium. The points should be separated positively, and be pressed together by means of a spring. Reversing the direction of the current through the points will prevent pitting. A condenser connected across the points is essential, but this is usually contained in the coil.

Cables and Connections.—Both high- and low-tension cables are best rubber-covered; and as nothing lends itself more easily to adulteration than rubber, it is advisable to buy cable from well-known makers only. The total diameter of copper strands should not be too small, both for mechanical as well as electrical reasons. It is most important that all contacts should be thoroughly clean and tight, and the cable should be so supported that there is no chance of the rubber being frayed against a metal edge by the vibration.

Dry Batteries.—These should be chosen with as large capacity as can conveniently be carried, not only for the reason that they will last longer, but chiefly because the internal resistance, which is an important item with dry batteries, is much less in the case of a dry cell of large capacity. The voltage of a dry cell is usually given as 1·5; but this is measured on open circuit, and when a good current is taken from the cell the pressure drops quickly. It requires four dry cells in series to properly take the place of a 4-volt accumulator.

Accumulators.—The proper charging of accumulators is an important matter. When a 4-volt accumulator shows only 3·6 volts on a voltmeter, it should be recharged at once. When fully charged the accumulator should certainly show 4·4 volts, and immediately after charging, or when being actually charged, it will show a higher pressure than this. With a new accumulator which has never been charged, very great care must be taken to follow out the instructions which are usually supplied with any well-known make. The acid must be of correct quality and density, and the battery must be charged electrically for a long time without stoppage, until both compartments are gassing freely.

The most convenient form of accumulators to use are those mounted in celluloid boxes, so that the plates can be seen. Some accumulators have the electrolyte in a jelly form. These do not give as good results as those having ordinary liquid electrolyte. When the accumulator is fully charged, the positive plates are chocolate in colour, and the negative plates grey. When discharged there is very little difference in appearance between the plates.

No acid should be allowed on the outside of the case, and the terminals should be kept very carefully clean. The terminals should be of such a design (usually made with rubber washers on) that the acid cannot creep on to the terminal. Where creeping does occur, thoroughly cleaning up the terminals and smearing all over with vaseline helps to prevent it. It is very important to always keep the level of the liquid above the tops of the plates. Much the best way to keep accumulators for any length of time is to have them in use ; but, where this is not possible, a short charge should be given every fortnight, but, in any case, the battery should not be left for more than two months without charging and should be fully charged before being left to stand idle.

Non-Trembler Coils.—These are very simple instruments, and practically the only important point is to see that they are properly fixed on the machine, all the terminals being clean and screwed up tight. A non-trembler coil should have a larger number of turns and thinner wire in the primary winding than a trembler coil. If this is not so, they are liable to take too large a current. The current taken may be reduced by inserting a resistance ; but this is a very inefficient plan, as it is far better to put the resistance into the circuit by means of more turns in the primary winding. When using non-trembler coils, it is very important always to switch off when stopping the engine, as, if the contact-breaking points happen to stop in the closed position a heavy current will go through the coil, damaging both it and the accumulators.

Trembler Coils.—The correct adjustment of the trembler makes a considerable difference to the running. The best way of adjusting coils is with special instruments for the purpose ; but this method of testing is hardly possible for the average motor cycle rider, and the next best plan is to adjust the trembler while the engine is running. It is impossible to give any definite instructions with reference to the setting of the tremblers, as there are so many different kinds ; but by carefully trying different adjustments, the best position will fairly easily be found. As a rule, an extremely light setting gives the best results at very high speeds, but will not give such good starting ; and so, as is usually the case, a compromise has to be effected. A good make of trembler coil takes about $\frac{1}{4}$ amp. for a single-cylinder engine, and about $\frac{1}{2}$ amp. for a twin-cylinder engine. If misfiring occurs, see that all the trembler parts are making a thoroughly clean contact, and are perfectly tight. It sometimes occurs that the platinum points come loose in their setting, and this is a difficult fault to detect. The platinum points should exactly coincide, and the surface of contact should be quite level. In trimming the contacts, only a dead smooth file should be used, and as little platinum as possible removed. Reversing the direction of the current through the platinum points prevents pitting, and is economical in platinum.

All coils want mounting in the machine in a place where they will be kept properly clean and free from moisture, and especially from any possibility of acid spray, or fumes from the accumulator. The coil should be firmly held so that it cannot jolt about.

Adjustments for a Twin-cylinder Engine.—These adjustments are best done with the engine running and, if possible, under load. When a double trembler coil is used, the tremblers should be set as near as possible to precisely the same position. A rough, though not definite, proof is when the two tremblers give the same note. On an engine in which the explosions occur at irregular intervals, the second firing cylinder requires to fire a shade earlier than would be the case if the explosions were exactly at equal intervals.

CHAPTER VIII.

THE TRANSMISSION : BELT AND CHAIN DRIVES, FREE ENGINE CLUTCHES, AND VARIABLE SPEED GEARS.

IN the preceding chapters we have seen how the motor is constructed, and have followed in succession the principle upon which it works, the production of the fuel it consumes, and the means and apparatus used in the process of ignition.

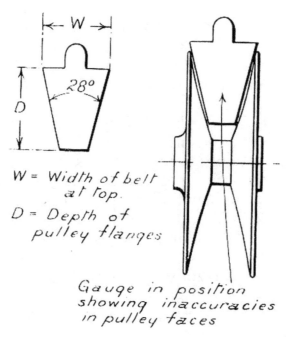

W = Width of belt at top.

D = Depth of pulley flanges

Gauge in position showing inaccuracies in pulley faces

FIG. 87.—Testing the Angle of the Pulley Faces by means of a Gauge Plate.

In other words, we have completed the study of our subject so far as the *production* of power is concerned, and may now turn our attention to the manner in which that power is applied to its purpose of driving the motor cycle.

Belts and Belt Fasteners.

The most common form of transmission on a motor cycle is that of a " V " belt, working in a similarly shaped pulley on the engine shaft and in a belt rim attached to the rear wheel of the bicycle. This simple method of driving permits of a comparatively narrow belt being employed, as a sufficient

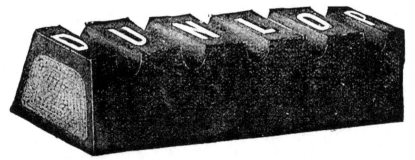

Fig. 88.—The "Dunlop" Rubber and Canvas Motor Cycle Belting

Fig. 89.—The "Severn" Rubber Belt.

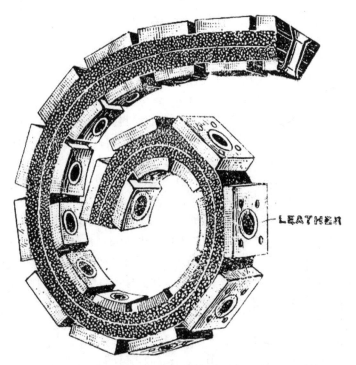

LEATHER

Fig. 90.—The "Service" All-Leather Belt.

grip can be obtained in the pulleys by its wedging action ; whereas if a flat belt were employed, as in workshop practice, a much wider one would be needed, and the likelihood of slipping would be greatly increased. The angle of the pulley flanges and that of the belt sides must, if a sufficient grip and full power output are to be secured, strictly coincide with one another at 28 degs., and the pulley groove must be sufficiently deep to preclude all possibility of the belt coming into contact with the bottom, for, if that occurs, there will be a considerable, and possibly a total, loss of power due to slipping. The pulley should be examined when it has been in use some time, to ascertain whether the flanges have worn to an incorrect angle. For testing this a gauge cut to the same width at the top as the belt, and having its sides a true 28 deg. angle, should be used in the manner illustrated in Fig. 87. Round, twisted belts have been used with success on light machines fitted with engines of $1\frac{1}{4}$ or $1\frac{1}{2}$ h.-p. in conjunction with jockey pulleys as tensioning devices. There is, however, a decided tendency to fit even these low-powered models with V belts in the latest practice.

Motor cycle belts are made in several forms—some being of rubber with a tough core of compressed canvas ; while others are made of leather strips or layers, riveted or otherwise connected together. Each kind has its advantages, the rubber belt being the cheaper and cleaner of the two, and easier perhaps to adjust ; but the leather belt lasts longer and grips better in wet weather. It is customary nowadays to mould rubber belts with transverse grooves on the underside to allow of their bending the more readily round the engine pulley ; while, in the case of one well-known make, small grooves are formed at the top also. The great idea is to make the belt as flexible as possible, and the purchaser of a belt should take it in both hands and see whether it will readily conform to a small radius such as it will have to adapt itself to on the engine pulley.

The author has met with very satisfactory experiences with the " Dunlop," " Lyso," and " Severn " rubber belts ; while of the leather ones he has used, the " Watawata " and " Service " belting have given by far the best results. The " Watawata " (Fig. 91) is an extremely flexible belt and its strength is enormous, while the " Service " (Fig. 90)—a slightly cheaper production—fully equals it on general grounds.

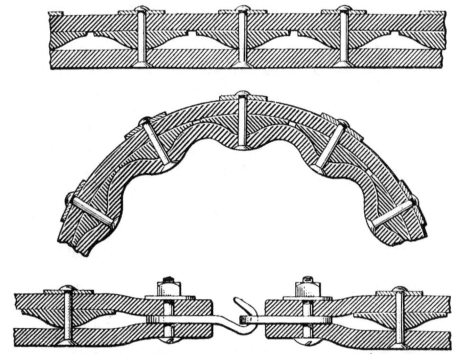

FIG. 91.—The " Watawata " Leather Motor Cycle Belt and Fastener.

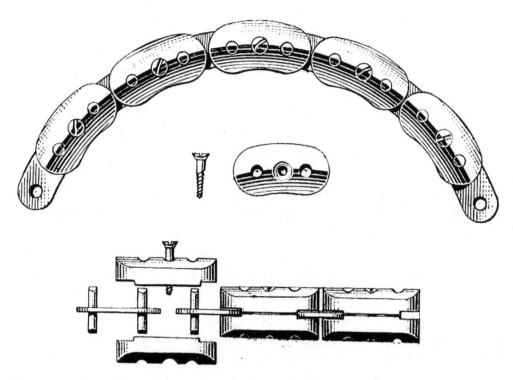

FIG. 92.—The " Whittle " Motor Cycle Belt; Steel Links with Leather Covering; no fastener required.

All belts stretch when first used, and it may be necessary to shorten them once or twice during the first hundred miles or so run to prevent slipping. A sharp knife (or small saw) and a bradawl or belt punch should be carried at all times, and by the aid of these appliances the task is one of but a few minutes. In the case of leather belts, it is usual to provide several eyelet holes or hollow rivets instead of solid ones for some distance from each end, so that when the necessity for shortening arises the rider is not confronted with the task of drilling out the hole for the fastener screw. It is unwise to go out for a long ride with a leather belt in which none of these openings are provided, for, despite what may be said to the contrary, it is usually a rare job to get rid of the obstructing rivet when out on the road and shortening has to be effected.

As to belt fasteners—their name is legion. The most useful ones are those which permit of the belt being removed from the machine without having to displace the fastener itself. These are called *detachable* fasteners and are made in two separate portions, engaging the one with the other by means of a hook or link, which only needs disconnecting to allow of the belt being immediately and entirely removed. The connecting hooks or links are made in two or three different lengths, so that shortening of the belt may be accomplished without having to cut it.; but, of course, when any really appreciable stretch has taken place, the last-named expedient must be resorted to. In the case of the " Whittle " belt (Fig. 92), which is virtually a steel chain encased in leather, no fasteners are required, shortening being effected by removing a complete section at a time. The leather pads are held in place by means of ordinary wood screws. This is the belt *par excellence* for the heaviest motor cycle work, and also for belt-driven cycle cars, and the fact of there being no fastener is a point greatly in its favour, especially with cycle cars with their double drive.

It is very important that belts for motor cycles shall be suitably proportioned for the work they have to do—say, $1\frac{1}{8}$ ins. wide at top for anything above a 5 h.-p. engine, 1 in. for engines between $3\frac{1}{2}$ h.-p. and 5 h.-p., and either $\frac{3}{4}$-in. or $\frac{7}{8}$-in. for all but the smallest engines (such as $1\frac{1}{2}$ h.-p.), which work very well with $\frac{5}{8}$-in. belts. It is also very important that the belt be maintained in good and flexible

FIG. 93.—The " Forward " Belt Fastener : Adjustable and Detachable.

FIG. 94.—The " Terry " Detachable Belt Fastener.

FIG. 95.—The " Stanley " Detachable Belt Fastener, with Adjusting Links.

FIG. 96.—The " Simplex " Detachable Belt Fastener.

condition, rubber ones being sponged over with water or petrol, and leather ones scraped and dressed with collan oil occasionally. The belt should be slipped off the pulley when not in use ; it should never be run any tighter than is necessary to guard against slipping, for a very tight belt absorbs power and may break. The pulleys should be kept as free as possible from accumulation of grit and grease, and in attaching the fastener care should be taken to place the screw as far back in the belt as circumstances will permit, and perfectly central. Above all, great care should be taken to ensure the belt being in true alignment—*i.e.*, both pulley

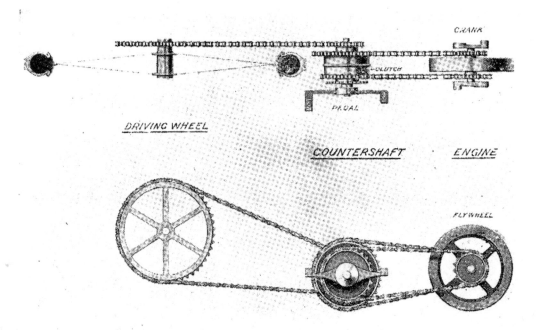

FIG. 97.—Chain Transmission of " Scott " Two-stroke Motor Bicycle.

and belt rim exactly in line with one another. Errors in this direction cause excessive wear of the belt, and, in aggravated cases, may bring about its constant shedding from the pulleys.

Chain Drive.

Chain-driven motor bicycles are less numerous by far than belt-driven ones. Still, some of the very best makers employ this method exclusively, and there are indications that it will become even more generally adopted in the future. Chain drive was considered to be *the* method of transmission for motor cycles in years gone by, and only because of the great improvements made in belt manufacture

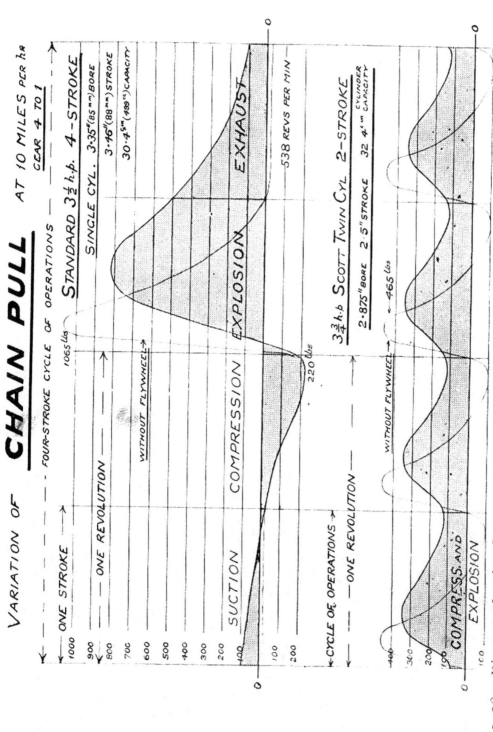

Fig. 98.—Diagram showing Comparative Performance of Four-Stroke and "Scott" Two-stroke Engines in transmitting power by chain.

was it ever superseded. It is customary to employ two chains—one engaging with the sprocket on the engine shaft, and also with a countershaft at the bottom bracket ; and the other extending rearward from the countershaft to the driving wheel of the bicycle ; this plan doing away with the need for a long chain with its attendant disadvantages of backlash and consequent likelihood of jumping the sprockets. Users of the belt method assert that chain-driven machines are more severe on tyres, less flexible to drive, and more prone to side-slip ; while the chains are even more messy to handle in muddy weather and more difficult to repair than belts, while they require greater care all round ; but, on the other hand, the advocates of the chain strenuously deny these allegations, and regard (or affect to regard) the belt users as somewhat behind the times and as poor misguided folk.

Motor cycles are nowadays sometimes fitted with what may be termed the combination type of transmission. In this system a chain is employed as between the engine and a countershaft, the latter occupying a position in the bottom bracket with belt drive from the pulley mounted on this countershaft to the rim on the back wheel. A two- or three-speed gear mechanism is, as a rule, carried in the bottom bracket position, and a pulley of large diameter is rendered possible—a fact which makes it possible to avoid what otherwise would be an objectionable feature, viz., an unusually short belt drive, involving a much increased risk of slipping of the belt. Many experienced riders favour this combined form of transmission which provides a positive drive in the first stage, whilst retaining the flexibility and other advantages of the belt between the pulley and the back wheel. As an example of this practice, the "Douglas" motor cycle, Fig. 181, p. 233, may be cited.

Shaft Drives.

A few makes of motor bicycle, notably the " F.N.," are driven by means of bevel-gear ; the engine shaft in this case running longitudinally instead of transversely to the frame and engaging with a cardan or propeller shaft, carrying at its rearward extremity a small bevel wheel which engages with a larger bevel wheel attached to the rear wheel of the bicycle. This plan is applied to both the four-cylinder and two-speed single-cylinder always given satisfaction. of transmission is practically weather - proof and, moreover is always working under the best conditions to resist wear, all frictional parts being covered with grease. The arrangement is illustrated in Figs. 99 and 102A.

The " T.M.C." four-cylinder motor bicycles have a similar arrangement, but a worm drive is employed instead of the bevel gear - wheels. This last-named machine, of which illustrations are given on pages 195-6, and 249, approximates in many directions to car practice in matters of control, and presents many features which strongly differ from those of any other type of motor cycles.

Adjustable Pulleys.

We may next proceed to a consideration of transmission developments permitting of varying the gear, the simplest and least expensive of all being the adjustable pulley. This is merely an ordinary

" F.N." motor bicycles, and has Being entirely enclosed, this type

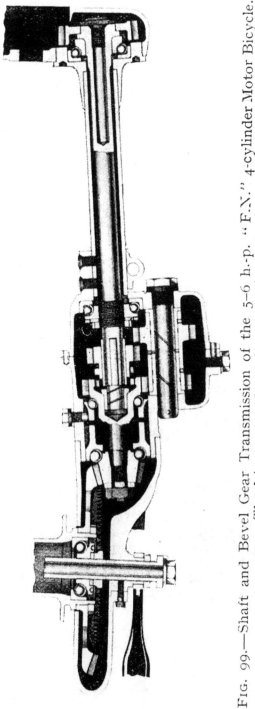

FIG. 99.—Shaft and Bevel Gear Transmission of the 5–6 h.-p. "F.N." 4-cylinder Motor Bicycle. The drive passes through a 2-speed gear-box.

V pulley fixed as usual on the engine shaft, but so devised that the outer flange may slide or rotate in such a manner relatively to the inner or fixed flange that the V groove may be widened and the belt caused to sink further into it.

It will readily be seen that when this occurs the relation between the diameter of the engine pulley and that of the belt rim on the back wheel of the bicycle will vary ; or to put it more technically, the gear ratio will be lowered. If the pulley has an outside diameter of 5 ins., and the belt rim a diameter of 20 ins., the proportions are as 4 to 1, and the machine is said to have a *gear* of 4 to 1 (the engine pulley making four revolutions to each single revolution of the driving wheel) ; but, when by manipulating his adjustable pulley the rider lets his belt sink down into the groove so as to pass round, say, a 4-in. instead of a 5-in. radius, as before, he alters this proportion and obtains a gear of 5 to 1, while if he further opens out the flanges he can get a lower gear still. Of course, it is only possible to vary the gear within somewhat narrow limits by this means, for, as has already been pointed out, under no circumstances must the belt be allowed to ride on the bottom of the pulley, while to make the latter much larger in diameter in order to gain a greater range of variation would only be to defeat the object aimed at, and render the higher gears impossible of use, except perhaps under the most favourable conditions of running on the flat. Adjustable pulleys make it necessary to employ some means of altering the length of belt to suit the varying conditions of gear, and this is where the detachable link

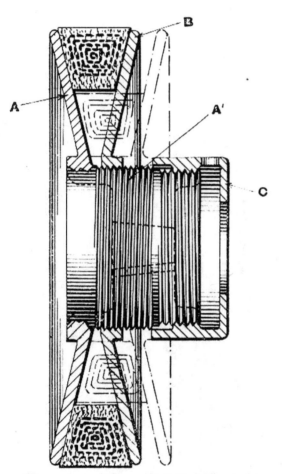

FIG. 100.—The "V.S." Adjustable Pulley.
(A)—Fixed Flange.   (B)—Movable Flange.   (A1)—Left-hand Thread.
(C)—Locking Cap.

fasteners come in very useful indeed. Another method is to cut the belt to the right length for a low gear and to make up the difference required for higher gears by inserting a short length of belt (a few inches only) with an additional fastener, and if the belt is adjusted rather on the tight side it will still grip when the gear is slightly lowered. The adjustable pulley arrangement is a very convenient one, as the gear can be lowered for hill-climbing and raised for level road work without—excepting only in the slightest degree—adding to the complication of the construction.

A simple example of adjustable pulley design is shown in Fig. 100. This type of pulley is fitted to the "V.S." motor cycles and has proved very efficient. The pulley flange (A) is formed in one piece with the boss portion (A¹), which is secured to the taper portion of the engine shaft by means of a feather and outside nut. The boss (A¹) is provided with a left-handed screwed part, and outside that is a right-handed screwed part, as seen. The outer flange (B) of the pulley rotates on the left-handed screwed portion so as to be capable of being moved either nearer or further away from the fixed flange, while the smaller screwed portion of the boss carries a loose cap (C), which serves to lock the movable flange in any desired position. The pressure of the belt (D) against the inside of the loose flange (B) while working, forces the flange outwards against the locking cap (C), so that there is no chance under any circumstances of the flange slipping and automatically lowering the gear by allowing the belt to sink lower into the groove. This pulley gives a range of gears from about $3\frac{3}{4}$–1 to $5\frac{1}{2}$–1. Fig. 101 shows the "Premier" adjustable pulley, and there are numerous other methods of arriving at the same result, but space prohibits their being illustrated or described in these pages.

FIG. 101.—The "Premier" Adjustable Pulley, similar to the "V.S."

Free Engine Clutches.

The next development connected with the transmission is the free engine clutch—a device for allowing the engine to continue in motion, though the bicycle, as a whole, remains stationary with both its wheels on the ground. A clutch of

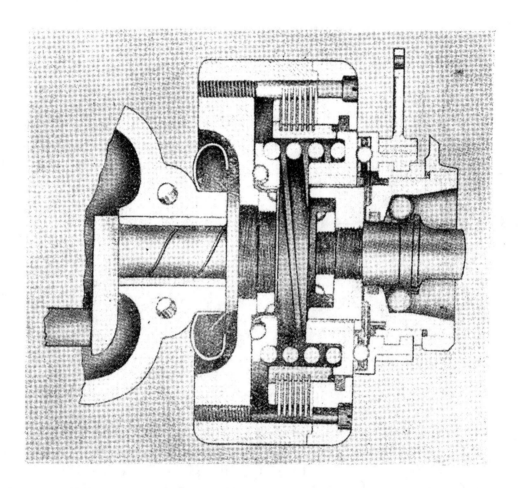

FIG. 102.—The Clutch Mechanism of the "F.N." 4-cylinder Motor Bicycle, with Shaft and Bevel-gear Transmission, is located in the Flywheel, as illustrated above. The clutch is manipulated by a small lever on the handle-bar; it is very sensitive and easily controlled. The multiple discs and powerful coil spring are plainly seen, as is also the crankshaft of the engine and its connection with the flywheel.

this description may either be attached to the engine shaft or contained in the hub of the back wheel, and in either case an adjustable pulley may be used as well. A free engine clutch is an undoubted boon to the motor cyclist, for, if nothing else, it does away altogether with any necessity to run alongside the machine to mount, while, in traffic round corners, and on hills, the control of the engine is vastly

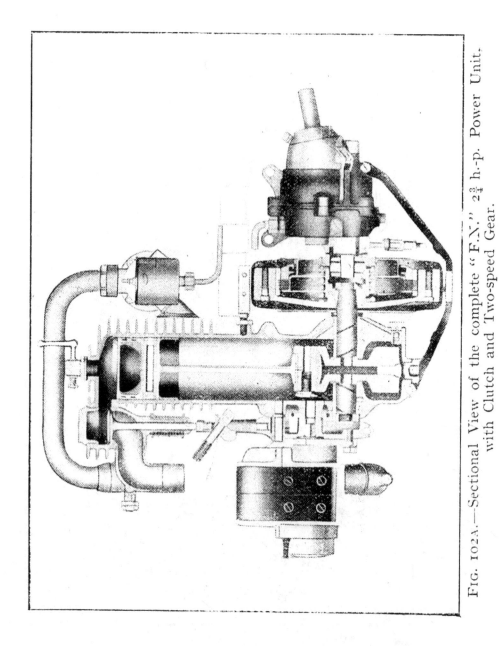

FIG. 102A.—Sectional View of the complete "F.N." $2\frac{3}{4}$ h.-p. Power Unit, with Clutch and Two-speed Gear.

facilitated. The author has had long experience of the Mabon clutch, and believes it to be the most efficient of any among those designed for attachment to the engine shaft.

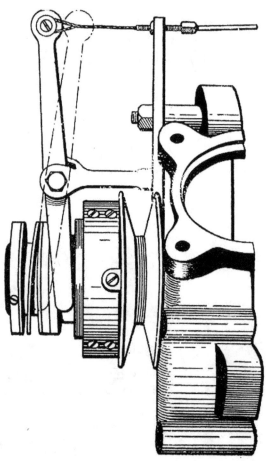

FIG. 103.—The "Mabon" Free Engine Clutch for Motor Cycles.

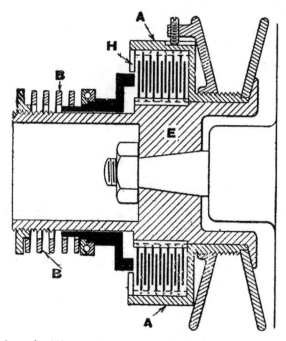

FIG. 104.—Sectional Elevation of "Mabon" Multiple Disc Clutch
and Free Engine Device.

The " Mabon " Clutch.

This device, of which sectional drawings are given herewith, combines a free engine adjustable pulley and metal-to-metal disc plate clutch, comprising a drum or box (A) which carries a series of discs or plates and is formed in one piece with the body of the pulley. The outer flange of the pulley is capable of being rotated on the screwed portion of the body to allow of the gear being raised or lowered, on the principle already described.

An inner body portion (E) carrying a second series of plates is secured to the engine shaft in the manner usual with an ordinary pulley, and it will be noticed that the plates carried by the drum (A) and those of the portion (E) just referred to, are arranged so as to interleave or follow one another, in alternate order.

A sleeve (blackened in Fig. 104) is caused to slide in and out of the drum (A) for a limited distance under the action of a strong coil spring (B), the movements of which latter are controlled by means of a forked lever (Fig. 103) pivoted on the end of a short transverse arm connected to a bracket which is bolted to the crank-case as shown. To the outer end of this lever there is attached a Bowden wire, which leads through the bracket (where an adjusting stop is provided) upwards to the handle-bar and there connects with a hand-lever controlled by the rider, this lever being provided with a spring stop so as to avoid the necessity of retaining a hold upon it when the clutch is out. It will thus be seen that the lever on the handle-bar actuates the Bowden wire, the wire moves the outward outer end of the lever backwards and forwards transversely, and this causes its forked end, influenced by the coil spring (B), to work the sleeve in or out of the drum as desired, thus in the first case bringing the two series of plates into frictional contact with one another, when the engine, of course, drives the machine forward in the usual manner, or releasing the plates from contact when the engine runs free without propelling the machine. Between the friction or driving position, and the loose or free engine position—that is to say, when the spring and sleeve are only partly withdrawn—it is possible to obtain almost any degree of slip required so that the engine, without being altogether released from its load, can be accelerated and its power thereby increased before again letting in the clutch and taking up the drive. In the latest design the forked lever is dispensed with in favour of the rotating sleeve arrangement shown in Fig. 105. This latest device does away with all suggestion of end-thrust, and is particularly sweet in its action.

Motor bicycles fitted with this device can be started from a standstill in the same manner as a car. The rider starts his engine by means of the pedals or otherwise, while the rear wheel is jacked up clear of the ground. He then pulls the lever on the handle-bar into the position necessary to

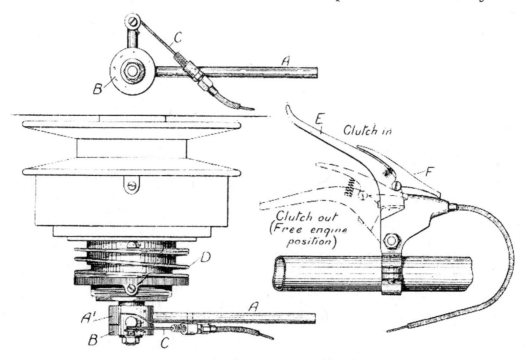

Fig. 105.—The Latest Development of the " Mabon " Clutch, showing Handle-bar Controlling Mechanism.

free the engine, lets the wheel down on to the ground, and sits astride the saddle, the machine being, of course, entirely motionless the while. Then by gradually engaging the clutch the drive is taken up by the engine and the machine moves slowly forward, gathering speed according to the whim of the rider and his manipulation of the clutch. The clutch-operating lever should always be gradually manipulated, for, if suddenly engaged, the engine will be unable to accommodate itself to the load and will come to a stop. If traffic of sufficient density to impede his progress confronts him, the rider, by disengaging the clutch, can bring the machine to a standstill, keeping the engine running, and re-engaging when an opportunity for going ahead presents itself. If obstructed on a hill and the speed of the engine falls below what is necessary to maintain the power required, it can be restored by allowing the engine to run partially or wholly free for a few seconds, and in a hundred different circumstances the clutch will be found useful in controlling the motor cycle. It is, perhaps, superfluous to add that

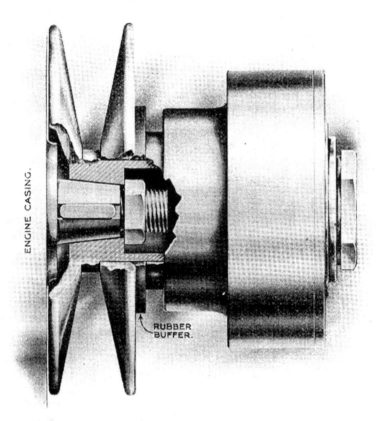

FIG. 106.—The "Philipson" Governor Pulley.

while sitting astride the stationary machine the rider maintains his balance by placing one foot on the ground.

The Philipson Governor Pulley.

The Philipson governor pulley for motor cycles is of recent introduction, and it appears to possess the elements necessary to ensure success; indeed, it has already proved efficient in open competition at hill climbs and other motor cycling road events. The construction is of a simple character, as seen in the drawings Figs. 106 and 107. The pulley is of the expanding type with the fixed or inner flange (A) secured on the tapered end of the engine shaft in the usual manner. With the flange (A) is incorporated a hollow sleeve or extension (B) which carries the sliding flange (C), a key-piece being provided as shown. The flange (C) also has an extension, and on the outer end of this latter there is a

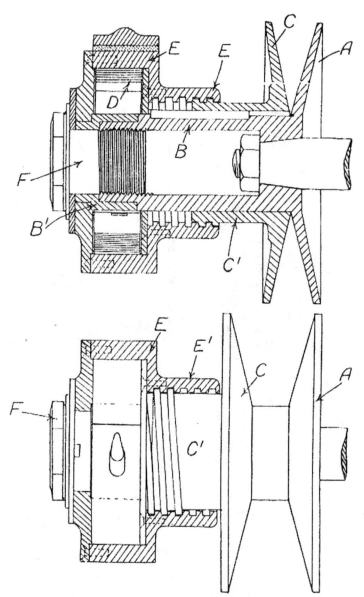

FIG. 107.—Views of the Philipson Governor Pulley.

coarse threaded screw. The spring box (E) has an internally threaded portion, the thread in which corresponds to that on the sleeve (C) and engages therewith.

Inside the box (E) is a powerful flat coiled spring, one end of which is secured to the inner surface of the box itself and the other end to a bush (B¹) in connection with the sleeve (B). One end of this bush (B¹) acts as an outer carrier for the spring box (E) by means of a cover or lid, this lid being secured to the spring box by two screws having countersunk heads.

The whole arrangement is locked in position by the screw (F) which passes through the bush (B¹) into the main sleeve (B). It tightens up the bush (B¹) but leaves the spring box (E) free to revolve, and in doing so it opens and closes the pulley flanges. The whole of the parts revolve at the same speed as the engine.

The belt is held under a constant grip between the flanges

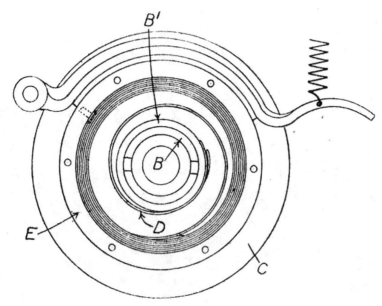

Fig. 107A.—Control Lever and Spring of the Philipson Pulley.

by the spring in the box (E), and when the speed of this box is retarded by outward pressure, either by the rider's foot or a pedal brake, the flanges are opened and the belt sinks, reducing the gear ratio accordingly. The retardation causes the spring to be wound up; therefore, when the retarding pressure is removed, the recoil action of the spring gradually and automatically forces the flanges into their normal or closed position, and the gear ratio is, as a consequence of this, raised. The maximum gear can be adjusted to anything between the limits of the flanges.

Hub Clutches.

Other free engine devices are those which are fitted in the hub of the rear wheel of the bicycle, and their manipulation is effected by means of levers placed within easy and convenient reach of the rider's foot. This system has the advantage, although not now a unique one, of allowing the engine to be started without the necessity of jacking up the rear wheel clear of the ground so that the rider may take his seat, press down one of the pedals sharply, and the engine will start, all being then in immediate readiness to go ahead by simply engaging the clutch. To permit of this, the belt rim

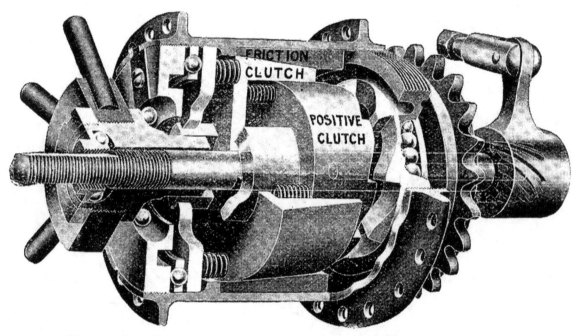

FIG. 108.—The " Villiers " Free Engine (Hub Type) Clutch.

is arranged to revolve independently of the wheel during such times as the clutch is " out," and the act of engaging the latter causes the rim and wheel to become locked and to revolve together.

The " Villiers " Free Engine Hub.

It will be seen from the illustration (Fig. 108) that there is a spoked drum for the driving belt at one end of this hub, and an independent free-wheel at the other. The clutches are the connecting media between these two. They consist of both a frictional clutch and a positive clutch operated by one foot control.

When starting from free engine position, the pressure of the heel on the foot control engages the frictional clutch, and transmits a gradual movement to the motor cycle.

When a normal speed has been obtained, the positive clutch can be let in by a further pressure of the heel, and slipping then becomes impossible.

To free the engine, the reverse operation is performed by a pressure of the toe on the foot lever, this first removes the positive clutch and then disengages the frictional clutch.

The springs shown in the accompanying illustration hold the clutches in engagement, they are compressed, and the

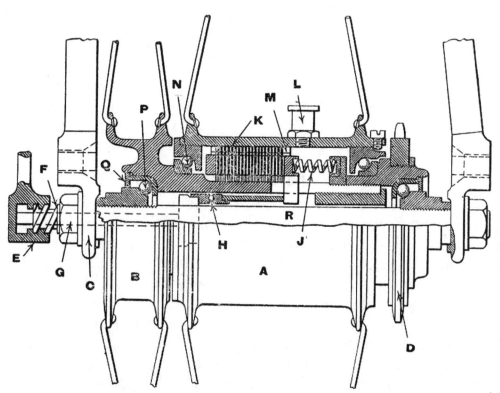

FIG. 109.—The "Rex" Free Engine Clutch, fitted in hub of rear wheel.

clutches disengaged by a rod running through the centre of the axle. This rod is actuated by a quick thread cap connected to the foot control.

The "Rex" Free Engine Clutch.

A good example of this type of clutch is shown in Fig. 109, which illustrates the "Rex" clutch fitted to the "tourist" pattern motor bicycles of that name. The two main sections are the outer shell, to which is attached the wheel, and the inner sleeve to which is coupled the belt pulley wheel. On the wheel section are eight feather keys, and these eight keys register with eight slots on the metal compensating external cone. The pulley section is also fitted with eight feather keys, on which slide two internal cones. These internal cones are pressed home concentrically by means of springs, thereby giving a rigid drive. A free engine is obtained on the coarse screw (F). This operation causes the plunger (G) to compress the springs (J), which act on clutch cones. It is operated by foot-pedals, similar to the "Rex" two-speed gear.

It will be noticed that the hub contains two ball races—namely, (N) the free engine ball race, and (P) the driving ball race. This is a great feature of the "Rex" hub and renders it exceptionally free in its action, and, by obviating friction, prevents undue wear and strain in the working parts of the hub. The engine is started when the machine is stationary by a downward thrust of the pedal.

The "Triumph" Clutch.

Another good example of the hub type of clutch is that of the Triumph Company, illustrations of which are appended. This clutch consists of a number of grooved steel plates, half of which engage on the axle and the other half in the ribs of the hub shell, these plates being held in compression by four springs. The drive is absolutely solid, and there is claimed to be no end thrust. When the clutch is out of action, the back wheel runs on a pair of ball-bearings, as in an ordinary hub. The clutch is operated by means of a rod, to which is attached a combined toe-and-heel pedal fitted to the footrest on the right side of the machine ; and in order to start the engine the rider (while seated in the saddle) puts the clutch out of action by pressing down the toe pedal, and next gives the crank pedal a smart push downwards. The clutch is then engaged by pressing down the heel pedal and allowing the toe pedal to gradually engage. The makers recommend the use of water-cooled oil for lubricating the clutch, and the hub should be periodically flushed with paraffin and well oiled afterwards. The weight added to the machine by fitting the clutch is about 10 lbs.

In other cases the engine is started by means of a handle

engaging with an extension on the axle of the rear wheel, and in this case also the belt rim revolves apart from the wheel itself. This system is known as the "live" axle system, and is incorporated with some of the most successful two-speed gears adapted to motor cycles.

It is a very difficult, if not an impossible, task to start the engine by means of a handle secured to the engine shaft, where magneto ignition is employed, owing to the fact that a sufficient speed of rotation cannot conveniently be got on

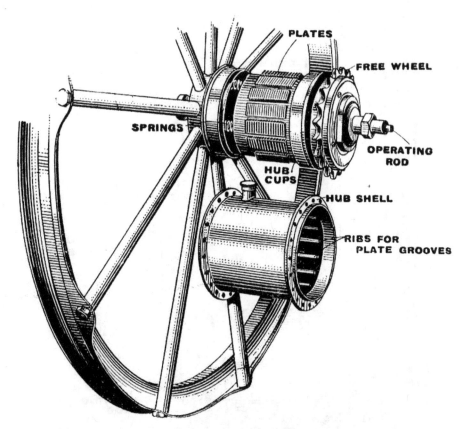

FIG. 110.—The "Triumph" Free Engine Device.

the armature of the magneto by this means to produce a spark of suitable intensity to fire the mixture ; but if the handle is affixed to a countershaft, or, as before described, to the live axle of the rear wheel, the extra leverage so obtained makes it an easy matter to set the parts revolving at a good speed before dropping the exhaust lever and setting the engine going.

Incidentally, a free engine clutch materially assists the

rider when wheeling his machine along, as he must sometimes do. He can then disengage the clutch, and by so doing stop the motion of the engine, so that there is no need to hold the exhaust lifter up to release compression, and both hands are left entirely free for handling the machine.

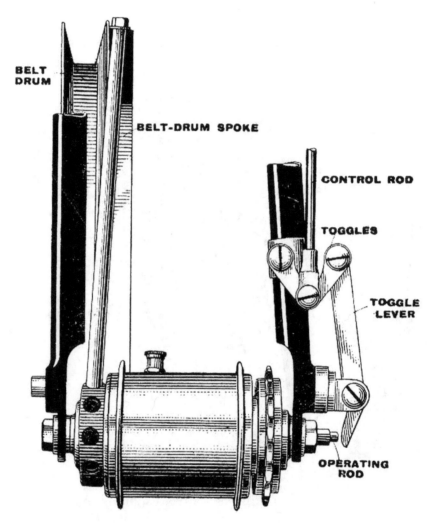

FIG. 111.—Hub Free Engine Clutch, "Triumph" Motor Cycles, showing Control Mechanism (in plan).

Variable Speed Gears.

We may next briefly consider the subject of variable speed gears for motor cycles. These are made in several forms, the majority affording a single change of gear, with a neutral or free engine position as well. These are known as two-speed gears in that they provide a high gear for fast running, under normal conditions, and a low gear for hill-climbing and for use in circumstances (such as when driving in thick traffic, and so on) which make it inconvenient to employ a high gear. Appliances of this kind are not confined to any particular class of machine, nor are they restricted for use with any one type of transmission alone. They are used alike with belt, chain, and shaft drives, and are fitted by some makers as a standard feature, while almost any

Fig. 112.—Starting a Motor Bicycle by the Free Engine Method. Engine running, but machine stationary. Rider steadies himself and machine until ready to engage the clutch.

motor bicycle can be equipped with some form of variable gear, and the plan is coming into greater favour every year.

Where the *frequent* carrying of an extra passenger under genuine touring conditions is contemplated, a variable gear becomes practically a necessity, although quite a large measure of satisfaction can be obtained under these circumstances even with a single-geared machine. The author, however, strongly recommends prospective " passenger "

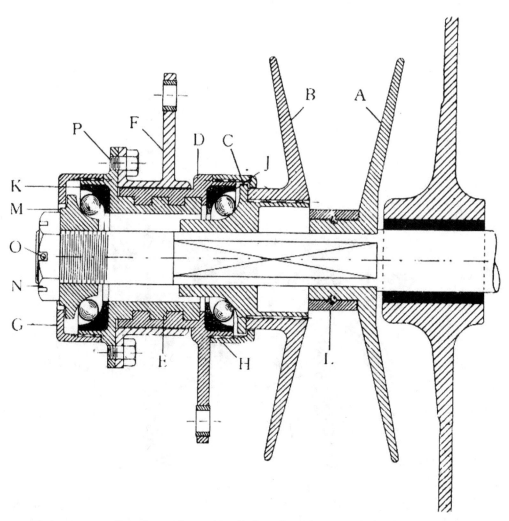

FIG. 113.—Section through "Gradua" Variable Speed Gear
(Expanding Pulley Type).

(A)—Fixed Inner Flange.
(B)—Sliding Flange.
(C)—Sliding Cone screwed in
Flange B,
(D)—Female Screw.
(E)—Male Screw.
(F)—Lever for top Gear-rod.
(G)—Outer Retaining Cap.
(H)—Inner Retaining Cap.

(J)—Ball Race in D.
(K)—Ball Race in E.
(L)—Ball-bearing Ring on
Flange A.
(M)—Left-hand Cone on Shaft.
(N)—Left-hand Lock-nut.
(O)—Split Pin.
(P)—Setscrews.

motor cyclists to invest in a variable geared machine or have their existing mount fitted with a device of this kind, failing which they should at least employ an adjustable pulley and free engine clutch.

Three-speed gears have now become very popular among motor cyclists, and they offer an undoubted advantage over those which afford one change of gear only. The Armstrong and Sturmey-Archer are the leading examples of three-speed gears, and both are being widely adopted by prominent motor cycle manufacturers.

The " Infinitely Variable " (or Expanding Pulley) Type.

Apart from these gears, there is another type known as the " infinitely variable " gear, by means of which the ratio

FIG. 114.—" Zenith " Motor Bicycle fitted with " Gradua " Expanding Pulley Gear Mechanism.

may be altered between the limits of the highest and lowest degrees by small and indefinite degrees, instead of, as is the case with a two-speed gear, only allowing of the one change —*i.e.*, from high to low. Thus, if the topmost gear is, say, 4 to 1, and the lowest 8 to 1, *any* intermediate ratio between these two points can be obtained by manipulation of the operating mechanism. The " Gradua " gear (Figs. 113 and 114) is an excellent example of this type, and the changes in the ratio are obtained by allowing the pulley to expand or contract and the belt to rise or fall in the groove, while

the back wheel of the bicycle is caused to move either backward or forward in the frame stays to maintain an even
belt tension. The gear is manipulated by means of a rotatable handle above the top tube of the frame, and the turning of this handle (which actuates bevel and chain-wheel
mechanism of a very simple kind) effects the movements
of the pulley and rear wheel above referred to. It will be
readily understood that when the belt is at the *top* of the

FIG. 115.—" Zenith " Kick Starter.

pulley the bicycle wheel is at the end of its forward travel
and the gear is the highest obtainable, and when the belt
has been allowed (by operating the controlling mechanism)
to sink down into the pulley as far as possible without coming
in contact with the bottom, the wheel is practically at the
end of its rearward travel, and the gear is the lowest obtainable ; while any intermediate variation provides at once a
slightly higher or lower gear as may be desired. The arrangement permits of the utmost flexibility being obtained from

the engine and almost any condition of running can be immediately provided for, however quickly changing it may be ; whereas with a two-speed gear it must be either the high or the low gear, and sometimes the latter has to be kept in use longer than is advantageous, because to jump up

FIG. 116.—The " Rudge Multi " Variable Gear Operating Mechanism.

from, say, 8 to 1 to 4 to 1 would, under certain conditions, cause the engine to labour, if not to stop altogether.

Free engine is obtained in the infinitely variable type of gear by allowing the pulley flanges to expand to such an extent that the belt comes in contact with the bottom where a ball-bearing slip ring is provided for it to ride upon, which it does without, of course, communicating the drive to the rear wheel of the bicycle, as it has no grip on the pulley flanges. Directly, however, the operating mechanism is brought into play, and the flanges are slightly contracted, or moved in towards one another, the belt at once grips and the drive is taken up on the lowest gear, and so on up to the highest gear, if desired, by continued manipulation of the handle.

The latest Zenith-Gradua machines are fitted, at the

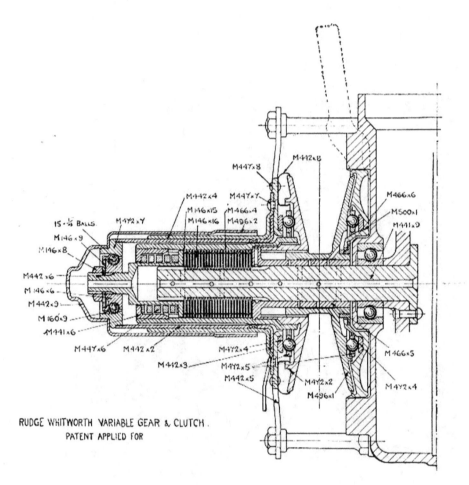

RUDGE WHITWORTH VARIABLE GEAR & CLUTCH.
PATENT APPLIED FOR

FIG. 116A.—The "Rudge Multi" Variable Gear Engine Pulley and Clutch Mechanism.

purchaser's option, with a kick starting arrangement (Fig. 115), this removing the objection which some motor cyclists have of being compelled to "push to start." In the absence of pedalling gear or other device adapted for the purpose, the whole machine must be pushed along in order to start the engine, the only alternative being the somewhat risky one of pulling the rear wheel round by hand against compression with the wheel jacked up off the ground.

With a cold engine, and a sidecar attached, this pushing of the machine entails some amount of physical exertion which many object to, and as it is impossible either to warm the engine up before starting, the task is rendered harder than it would otherwise be. With a kick starter available, however, this objection is removed, and with this refinement added, the Zenith-Gradua becomes one of the most easily managed and delightful machines on the market.

An Albion plate clutch hub is built into the wheel, and on a plate attached to the right-hand side of the axle are carried the brake and the handle starting arrangement.

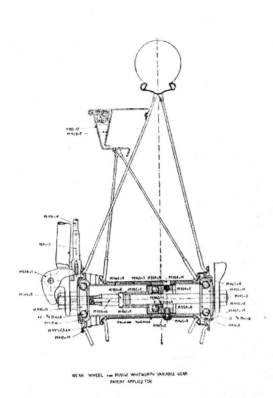

REAR WHEEL for RUDGE WHITWORTH VARIABLE GEAR
PATENT APPLIED FOR

FIG. 116B.—The " Rudge Multi " Gear. Arrangement of rear wheel hub and expanding belt rim.

This plate slides through a bracket attached to the frame, so that the brake and starter move backwards and forwards with the wheel. A fixed sprocket is screwed on to the back hub on the right-hand side, and a free wheel sprocket is attached to the spindle of the starting handle. This spindle is carried in an eccentric for the adjustment of the chain. The clutch is controlled with a pedal operated from the right-hand footrest ; a small pawl on the top of the pedal keeps the clutch in free engine position, so that when the foot is placed upon the pedal this pawl is held out of action. Free engine is obtainable on all gears.

The outstanding feature of the Rudge-Multi speed gear, illustrated in Figs. 116, 116A and 116B, is the maintenance of constant belt tension and perfect alignment of the belt on all gears. This is obtained by moving the outer flange of the back wheel pulley and the inner flange of the engine pulley in the same direction ; therefore, the belt line moves over bodily, but the grooves of both pulleys keep opposite one another and consequently the belt in perfect alignment. The drawings show the construction of the expanding pulley and clutch mechanism on the engine shaft (Fig. 116A) and the expanding belt rim and other mechanism carried by the back wheel (Fig. 116B).

The actual operation of the gear is effected by the movement of a long lever conveniently placed at the left-hand side

FIG. 117.—The "V.S." Two-speed Gear and Free Engine (mounted in rear wheel hub).

of the tank. The lever is held in place by the engagement of a spring catch in a series of notches—20 in number— thus there are twenty definite variations in the gear obtainable, and obviously it is only a question of providing more notches in the quadrant in order to secure as many changes of gear as might be desired.

In the standard machines the extreme limits of ratio provided by the gear are $3\frac{1}{2}$ to 1 and $7\frac{1}{2}$ to 1, a range sufficiently extensive for all ordinary purposes.

The operating lever is attached to a cam plate, which engages with corresponding cams on the face of the crank-case. On pushing this lever forward the cam plate with a sliding pulley flange attached is forced outwards by means of the cams on the face of the crank-case, thus closing the pulley and giving the high gear position. The reason for

FIG. 118.—Operating Lever, " V.S." Two-speed Gear.

making the inner flange the movable one is that the outer flange is connected to the plate clutch, which, according to standard Rudge practice, is situated on the engine shaft instead of in the usual alternative position in the rear hub, the makers claiming that the higher speed of the engine shaft gives a much more efficient and far more flexible clutch than those fitted in the back wheel hub, besides being far lighter.

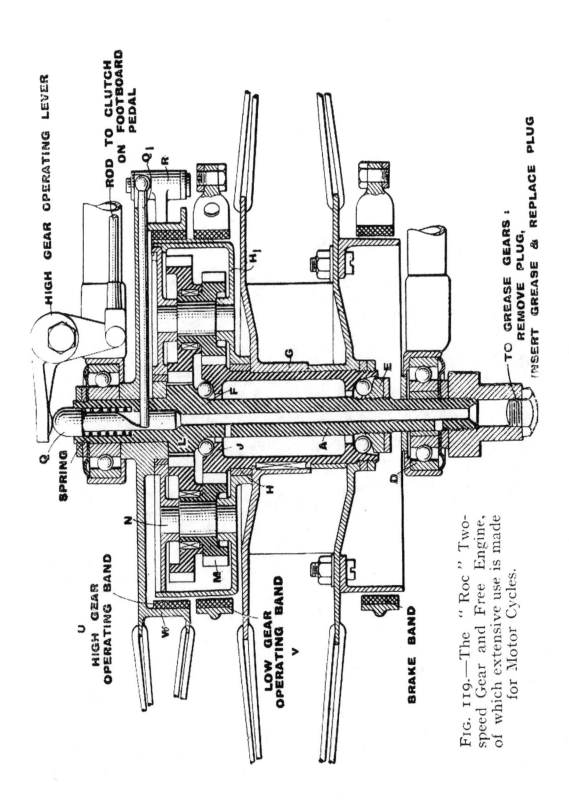

FIG. 119.—The "Roc" Two-speed Gear and Free Engine, of which extensive use is made for Motor Cycles.

At the same time that the operating lever is closing the engine pulley, a connecting-rod which is coupled to a bell crank lever (Fig. 116B) is performing a similar duty to the outer flange of the belt rim, only, as previously stated, both flanges move in the same direction, and as the inner flange of the engine pulley has moved outwards and closed that pulley, it follows that the outer flange of the belt rim is being moved outwards and therefore opens the belt rim and so allows the belt to drop to the bottom effective diameter of the rim.

As regards the actual operation of the gear, only a slight pressure on the lever is necessary. The change is effected at once, and, of course, no slipping of the clutch or closing of the throttle is necessary.

There are twenty-four pairs of hard steel plates in the clutch, oiled automatically through the centre of the axle. One of each pair engages with the axle and the other with the pulley. When pressed together by a powerful spring the plates cannot slip past each other, and the clutch drives solid. In this arrangement an expanding and contracting action in the belt rim on the back wheel is utilised instead of a backward and forward motion of the wheel itself, as in the Zenith-Gradua; but the effect is, of course, the same, namely, that of taking up belt slack as the gear is lowered. It will be realised by those who study the point that some provision of this kind must of necessity be made in the infinitely variable change-speed type of gear mechanism, otherwise the effect would be to produce slipping of the belt and consequent loss of power whenever the gear was lowered by means of expanding the pulley. The Rudge-Multi mechanism has proved highly successful on the road, and a very large number of machines so equipped are in use, some of them fitted with the large type of " Rudge " engine, which is rated as of 5 to 6 h.-p., whilst still retaining a single cylinder.

Two-speed Gears : The " V.S."

Of the two-speed motor cycle gears at present on the market, one of the most successful is that known as the V.S. This gear consists of planet pinions operated in such a manner that when one of each of the two sets of pinion wheels is held back by the brake designed for the purpose, the pair of wheels which carry the hub shell rotate round the pinion wheels. When running upon the high gear the hub shell is connected with the driving wheel of the bicycle by a toothed clutch : at such times—which, of course, greatly predominate—all gears are entirely out of operation. The gear is actuated by means of a lever placed on the left-hand side of the tank, with a quadrant to guide it. The central position gives free engine ; forward position high gear ; and

Fig. 120.—Collective View of the " Roc " Gear and Operating Mechanism.

backward position low gear. The method of starting and controlling the gear is very simple. The engine is first of all started by a push-down of the pedal and the lever placed in central (or free engine) position. Then all being in readiness for the actual start (the rider having taken his seat), the control lever is pulled slowly backwards until the brake engaging the pinion gears begins to grip. The machine will then start off slowly, but will pick up speed if the lever is pulled still further rearwards. The top speed is put into operation by slowly pushing the lever into extreme forward position, and it is advisable when doing this to lift the exhaust valve so as to reduce the power of the engine.

Immediately, however, the clutch is engaged the exhaust lever can be dropped, and the machine will then run on top speed in the same way as does the ordinary direct drive motor cycle. The reduction on the low speed is 50 per cent.

The " Roc " Gear.

The " Roc " two-speed gear and free engine clutch with live axle and handle starting has been developed by the makers, Messrs. The Roc Gear Co., Ltd., Birmingham, to a high pitch of excellence. They not only fit it to the motor bicycles of their own manufacture, but also supply complete " conversion " sets adaptable to almost any standard machine. The gear is constructed on the epicyclic principle— that is to say, with small pinions actuating central driving and driven pinions, with which they engage the gears; in fact, being always in mesh.

Referring to the accompanying drawing (Fig. 119), the live back axle (A) runs in bearings (D) held in the frame. At one end of the axle the fixed cone (F) is attached, and on the other end there is an adjustable cone (E) which gives all adjustments to the hub, thus making the gear and hub ball-bearing throughout. At one end of the hub (G) is attached the hub pinion (J). Behind this pinion is the phosphor-bronze bush (H) on which runs one side of the gear-box (H1). The belt rim centre (W) is built up on the live axle (A) which axle also forms the driving gear (L). The belt rim centre (W) is formed of two pressed steel plates, and riveted to this is a driving plate which meshes into the gear (L) which is fixed to the spindle. Inside the gear-box are three pinions (M) set equidistant from the gear attached to the spindle of the hub, and these pinions run on a hardened and ground steel pin (N), and through the centre of this the gear-box is clamped together by the bolt (O).

On the outside of the gear-box are two band brakes—one being attached to the belt rim (W), and the other (V) is attached to the frame. The method of operating the band brake (U) is as follows : One end of the live axle is bored to receive a taper push-piece (Q), which in turn actuates the rod (Q1) operating the lever (R) connected to the band brake. When the spindle is rotated by means of the starting

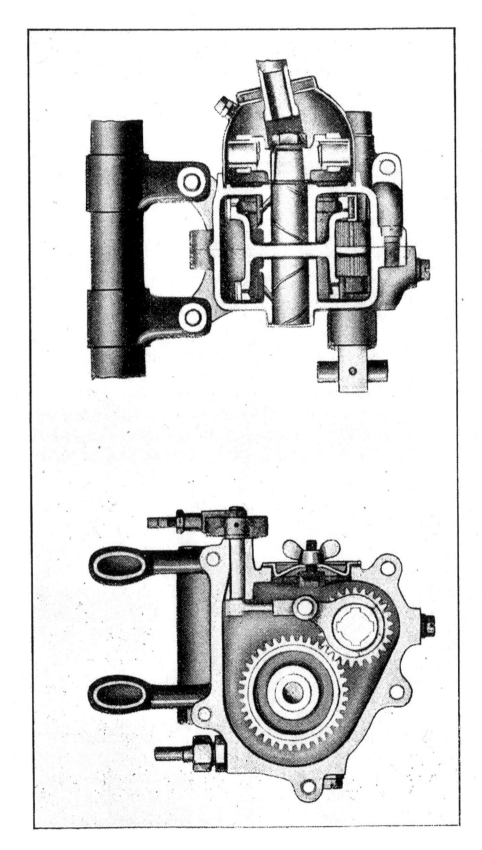

FIG. 121.—Two-speed Gear Mechanism fitted to F.N. Single-cylinder Machines.

handle it carries the belt rim round with it in a forward direction, giving the engine pulley several turns necessary to start the engine. The gear-box then revolves in the opposite direction to the belt rim through the action of the pinions. When pressure is applied to the band brake (V), it holds the gear-box stationary to the driving member; the low gear is then in action. On the other hand, when pressure is applied to the push-piece (Q) it tightens the band brake (U)—which is a fixture to the belt rim—on to the gear-box, thus driving *en bloc*, and the top gear is obtained. It will be noticed that when either the high or the low gear is at work there

FIG. 122.—The " P. & M." (Phelon & Moore) Two-speed Gear.

is practically no friction taking place in the hub. In the latest adaptation of the Roc gear for 1913 the hub and gear box are made in one piece, and single instead of double planet pinions are employed. The gear is also lighter and stronger, but the principle and main features remain the same.

When starting a motor cycle fitted with this arrangement, it is only necessary to raise the exhaust valve lifter and then rotate the live axle of the back wheel a few revolutions, dropping the exhaust lifter in the ordinary way when the engine is revolving at a fair speed. Then it is an easy matter to engage the low gear and move away slowly in a similar manner to a car, putting in the higher gear at the first opportunity.

The two-speed gear incorporated with the shaft and bevel drive of the F.N. 4-cylinder motor bicycle (Fig. 99) is of the sliding type with dog clutch engagement for the high gear giving a direct drive. The low gear is obtained by sliding a gear wheel on the primary shaft into mesh with a gear on the secondary shaft. The gear-box as seen is located at the rear of the transmission shaft casing ; indeed, it forms part of this casing. It is fitted with a quickly detachable inspection door, held *in situ* by a fold-over spring clip.

The operating lever is placed within easy access of the rider and has a suitable locking device for the respective positions of high gear, low gear, and neutral.

Another two-speed arrangement fitted to F N. motor cycles is shown in Fig. 121. This type is used on the single cylinder machines marketed by these makers. The gear-box is placed behind the engine and in a direct line with it. It is made in two halves, with the joints in a vertical direction. In order to obviate any possible tendency to excessive friction either on the shafts, their bearings, or the gear wheels themselves, due to the temporary distortion of the shafts when running over rough roads, a special type of universal joint is interposed between the clutch and the gear-box. The change-speed gear mechanism has only one travelling double pinion ; the multiplication of which, combined with the multiplication of the set of pinions which put the hub in gear, allows a demultiplication of the rear wheel in proportion to the engine of 1 to 6 for the high gear and 1 to 10 for the low gear. The two-gear shafts are disposed one above the other, and here again long bearings are employed. The gear wheels themselves are made of finest steel and accurately finished.

The " P. & M." Gear.

Another well-known motor cycle two-speed gear is the Phelon & Moore, fitted as standard on the machines manufactured by the firm of that name (which machines are usually referred to as the " P. & M.," for short). This gear is adapted for chain drive, as seen in the drawing. Briefly the mechanism consists of expanding rings made of bronze which work within hardened steel rings carefully ground to shape. The rings are expanded by wedge bars in combination with rollers suitably placed and operated by a ball disc

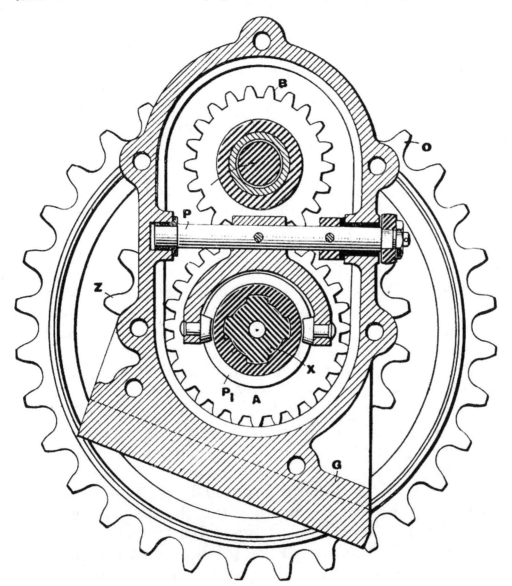

FIG. 123.—The "Indian" Two-speed Gear and Free Engine Clutch. Cross Section (through Gear-box on line *a-b*, Fig. 124).

bearing and fixed through a rod which passes through the main axle, and upon which the whole of the two-speed gear revolves. Cavities are provided in the rings for the storage of grease for lubrication and, once charged, this will last for a long time. As before said, the transmission is by chains, of which there are three—viz., high gear chain from the sprocket on the engine shaft to another sprocket on the countershaft, a similar arrangement for the low gear, and finally a chain drive from the countershaft to the rear wheel of the bicycle. The drawing (Fig. 122) shows the construction of the gear very plainly. The spring bronze rings are expanded by the movement of the wedge bars. When the thickest part of the wedge bars lies between the rings neither ring is expanded, and, therefore, both the gear-wheels run free instead of with the body of the gear. As the small sprocket driving to the back wheel is part of the gear itself, it follows that when neither of the gear-wheels is gripped by the expansion of the rings, the power of the engine is not communicated to the driving wheel. It is also obvious that the movement of the wedge bars in either direction will cause one of the bronze rings to expand and transmit the engine power to the back wheel, the drive being direct on both gears. The solid spindle operates the wedge bars, the thrust being taken by a double ball-bearing. Provision is made for starting the engine by handle on the countershaft, the engine running free until the low gear is engaged. The gear ratios of the " P. & M." motor bicycles are $4\frac{1}{2}$ to 1 (high) and 8 to 1 (low) ; while for those who desire to use their machines for taking a side car the gears are 5 to 1 and 9 to 1. The drive is taken up very gradually with this type of gear owing to the sliding action of the wedge bars, and, unless the rider is very careless indeed, he cannot very well injure any part of the mechanism.

The " P. & M." gear has been applied to certain other makes of motor cycles, and in one case the countershaft is placed forward of the engine and carries a belt pulley, in addition to the sprocket, the final drive to the rear wheel being by belt, thus combining the two systems of transmission.

In the " Indian " two-speed gear and free engine clutch, illustrated on pages 169 and 171, in section, the gear-box is fitted between the engine and rear wheel and occupies the position of the usual bottom bracket. The whole box can be made to slide on the frame, so that the engine chain can be easily and quickly adjusted. The clutch is a development of the compensating sprocket which has always been fitted on Indian motor cycles, and acts as a shock absorber to the transmission.

Referring to the illustration, the four pinions (A, B, C, and D) are always in mesh. Gears (B and D) are solid together and bushed to run on a stationary lay shaft, which is firmly keyed into steel bushings pressed into the aluminium gear case (G). The engine drives by chain on to the loose

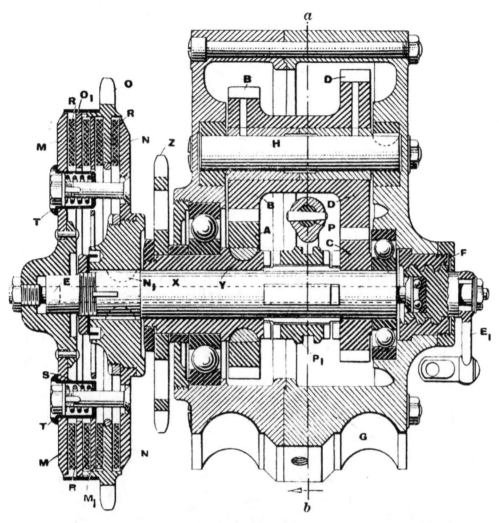

FIG. 124.—The "Indian" Two-speed Gear and Free Engine Clutch. Sectional Elevation. (Clutch on left, Gear-box on right.)

sprocket (O), and when the clutch is in the drive is communicated to the main shaft (X), the central portion of which is castellated and upon which slides the dog clutch (P1). This clutch has four heavy teeth on each face, corresponding with similar teeth on the loose pinion (C) and the loose sleeve (Y). Clutch (P1) is operated by forks on the shaft (P), which latter extends to the outer case and is connected by rods and levers with a change speed handle on the top frame tube.

It is evident that if clutch (P1) is moved to the left it locks shaft (X) to the loose sleeve (Y), causing a direct drive from the sprocket (O) to the sprocket (Z), which latter is firmly locked to (Y). The final drive is by chain from the sprocket (Z) to the rear wheel.

If, on the other hand, clutch (P1) is moved to the right to engage the pinion (C), the drive will be from the sprocket (O) on to the main shaft (X), through the clutch (P1) to the pinions (C and D) on the countershaft, thence by means of pinions (B and A)—which latter is firmly keyed to (Y)—to the sprocket (Z).

The neutral position is found by placing the dog clutch (P1) midway between the pinion (C) and the sleeve (Y).

It will be noticed that ball-bearings of ample dimensions are used wherever necessary.

The clutch is of the multiple disc type, the frictional surfaces in this case being respectively steel and Raybestos. The engine drives by chain on to the sprocket ring (O), which latter is free to rotate around shaft (X), as it runs on small rollers carried by the back plate (N), (N) being firmly fixed to (X) by the keys (N1). (O) carries with it the plate ring (O1) by means of dogs which lock them together. The plates (M M1 and N) are Raybestos faced and rotate with the mainshaft (X), as they are all three held together by pins (not shown in drawing). The plate (M), however, is free to slide endways on the shaft (X), and is normally forced towards (N) by the four springs (S) in the thimbles (T). In this position the clutch is in, and the rings (O, O1) are firmly held to (M, M1) and (N) by the friction between their surfaces, the drive passing to the main shaft (X) and thence into gear-box.

The releasing of the clutch is accomplished by a pin that slides inside the main shaft and is operated by the coarse pitch screw (F), which in turn is operated by the lever (E) ; this lever being connected to a foot lever conveniently situated on the left-hand side of the motor cycle. This pin butts against the adjusting screw (E) and separates the plates (M and N), thus freeing the gear ring (O) and allowing the engine to run idle. This arrangement enables the rider to set the clutch for any amount of slip he desires, and it will remain in the position adjusted. The clutch is easily accessible and easily adjusted, and all end thrust is eliminated while in the driving position.

FIG. 125.—The "N.S.U." Two-speed Gear (Engine Shaft Type). Adaptable to Standard Single Geared Machines.

Engine Shaft Two-speed Gears.

In addition to the gears already described, there are others—such as the N.S.U., Fit-All, and Osborne—which are carried on the engine shaft, the last-named providing

four speeds and a free engine. They are adaptable to almost
any make of motor bicycle, and require nothing more in the
way of fitting than the average mechanic is easily capable
of. Speaking from personal experience, the author can
cordially recommend the N.S.U. gear on account of its
simplicity, compactness, and certainty of action, and where,
as often happens, it is desired to convert an existing single-

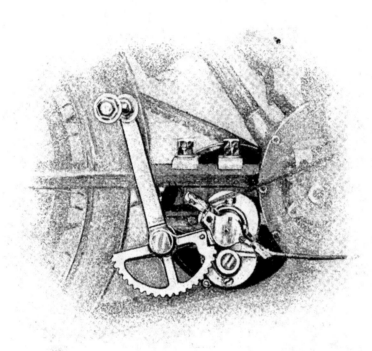

FIG. 126.—The "Bowden" Two-speed Gear (Countershaft Type).

geared machine to the two-speed principle, this gear offers
substantial advantages.

The gear is constructed on the sun and planet system,
in which several small spur wheels work between and in mesh
with a large internally-toothed ring and a small externally-
toothed wheel, so that the holding or releasing of the centre
wheel, the internally-toothed ring, or the ring holding the
intermediate wheels alternately, gives the necessary changes
in gear. In this case, the low gear is obtained by holding
the central wheel fast, and driving the internally-toothed
ring, fixed to the motor axle, thus driving the intermediate
wheels, fixed to the belt pulley, round the central wheel at a
speed 35% less than the speed of the motor.

At the bottom of the illustration the belt pulley can be seen, and also the ball races on which the apparatus runs when in the free engine position. The sun wheel of the epicycle gear is shown in cut, and the two wheels, shown in full at both sides, are the planet wheels. Further to the top the metal-to-metal clutch can easily be distinguished, also the clutch spring.

The N.S.U. two-speed gear is claimed to be the smallest and lightest two-speeder on the market, the weight complete being only about 9 lbs., and the over-all width about 5 ins. The easiest method of starting a machine fitted with the N.S.U. Two-speed Gear is as follows :—Place machine on stand, and start up the engine whilst the apparatus is on the high gear. Then throw out the clutch, and apply the back wheel brake so as to make sure that the engine is running free, and fold up the stand. Finally accelerate the engine by opening the throttle and advancing the spark, and turn the handle of the operating lever gently to the right, when the machine will take up the drive.

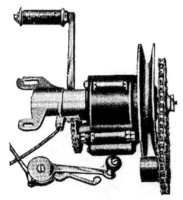

FIG. 127.—Another View of the " Bowden " Gear Mechanism.

The Bowden two-speed gear (Figs. 126 to 127A) is a good example of what is known as the " bottom bracket or countershaft " type. It is designed to fit behind the bottom bracket as shown in Fig. 127. Transmission is by ⅜-in. Renold chain from the engine shaft, on which a sprocket is keyed, in place of the ordinary pulley, to the gear-box and thence by belt from an 8-in. diameter pulley to the rear wheel. The drive is direct on the high gear, and, as seen, the belt and chain are both on the same side which tends to do away with cross strain and loss of power on the high gear. The sectional drawing (Fig. 127A), with its indexed references, clearly shows the constructional features of the apparatus. The two gears are obtained by expanding either pair of clutches (CC) into engagement with the recessed bores of the gear pinions (G and D) by means of the wedge bar (JJ) which operates through the centre of the main shaft.

Control of the gear is effected by means of a large ratchet

lever and Bowden wire mechanism from the handle-bar, or
by foot-lever working independently or in conjunction one
with the other. With the gear in free position the engine
may be started from the saddle with the road wheel on the
ground by a backward thrust of the kick starter.

Three-speed Motor Cycle Gears.

Of motor cycle three-speed gears, the Armstrong and

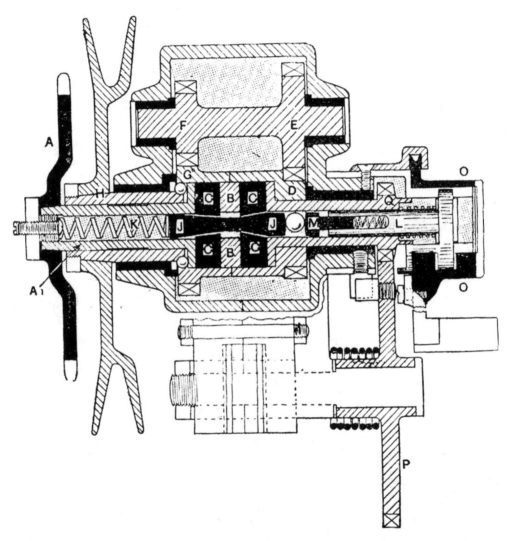

FIG. 127A.—Cross Section through "Bowden" Two-speed Gear
Mechanism.

Sturmey-Archer are the best known. Both are adaptable
to practically all makes of belt or chain-driven machines,
and both are being widely adopted by prominent manufacturers of motor cycles.

The recognised position for carrying the gear mechanism

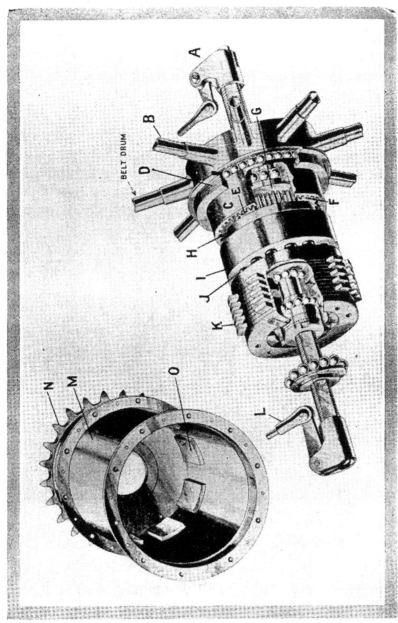

FIG. 128. The Armstrong "Mark V" Three-speed Gear (for Motor Cycles of medium horse-power).

(A)—Gear operating mechanism.
(B)—Belt Rim Spokes.
(C)—Internally toothed lower Gear ring.
(D)—Thread on to which Hub Shell is screwed.
(EF)—Planet Wheels.
(G)—Sun Wheel.
(H)—Carrier holding eight Planet Wheels.

(I)—Top-speed Gear Ring.
(J)—Clutch Springs.
(K)—Disc Clutch.
(L)—Clutch operating mechanism.
(M)—Hub Shell.
(N)—Free Wheel Sprocket.
(O)—Square Studs which fit the Driving Discs of the Plate Clutch.

is in the hub of the bicycle driving wheel, but quite recently
one firm has commenced to fit the Armstrong gear in the
bottom bracket position. In addition to this a Wolver-
hampton firm of motor cycle manufacturers fit their own
three-speed gear in this position. The Armstrong gear is made
in two patterns—Mark V, which drives " solid " or direct on
the middle gear ; and Mark VI, which is direct on the top gear.

The construction of each type—which differ only in detail
and not in principle—is shown in Figs. 128 and 129. In both
cases the engine can be started with the driving wheel of the
bicycle on the ground ; the gears are always in mesh and
run in an oil bath. The principle is the well-known sun and

FIG. 129.—The Armstrong " Mark VI " Three-speed Gear.

planet action, and a wide reduction of gear ratios is afforded.
A free engine clutch is incorporated with the gear.

The " Sturmey-Archer " Three-speed Gear.

This, like the Armstrong, is a hub pattern gear, and has
in a similar fashion been adapted for motor cycle conditions
after a long and successful career in connection with pedal
cycles. The hub is on the epicyclic principle and gives
three gears, with a free engine multiple plate clutch available
on each gear. Two epicyclic trains are used ; on the top
gear these are both locked and drive solid. On the second
gear one only is locked and the gear is reduced through one
train. On the low gear both trains are in action and a double
reduction is obtained. Choice of gear is determined by the
position of the sliding pinions (119 and 20), see Fig. 132,

which may revolve independently but are always moved together longitudinally. This movement is effected by a hand lever and quadrant fixed on left side of top tube of machine. The clutch is operated by a pedal on the right side, near the engine. The illustrations explain the operations.

The drive is transmitted by the belt drum to the driver (14), which is provided with internal gear teeth gearing with a set of planet pinions (19). These are mounted on axles forming part of the compound cage (25), which also has internal gear teeth gearing with a second set of planet pinions (26), mounted on axles forming part of the right-hand cage (27). Both cages are provided with clutch teeth which engage with the teeth on sliding pinions for the middle and high gears.

In Fig. 1 the high gear is engaged; the sliding pinions

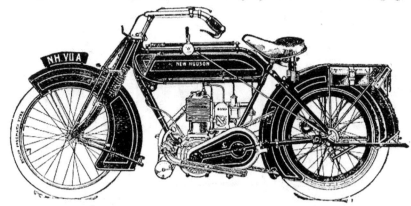

FIG. 130.—" New Hudson " Motor Bicycle fitted with Armstrong Three-speed Gear.

are locked to both cages, rendering the gear trains inoperative, and causing the right-hand cage to travel at the same speed as the driver.

For the middle gear the sliding pinions are moved mid-way to the left (Fig. 2). The large sun pinion (20) is then locked to clutch teeth formed in the internal clutch cone (101) which is held from rotation. This brings the left-hand train of gears into action and causes the compound cage (25) to travel at a lower speed than the driver. The small sliding pinion (119), is still locked to the right-hand cage (27), consequently the latter is revolved at the same speed as the compound cage.

For the low gear the sliding pinions are moved to the extreme left (Fig. 3). The large sliding pinion remains locked to the clutch cone, the small pinion is released from the right-hand cage, and the teeth on the small end of the pinion engage with a second set of clutch teeth formed in the clutch cone. This brings both trains of gears into action, and the right-hand cage is driven at a still slower speed.

The right-hand cage engages at all times with four lugs formed on the clutch body. This in turn drives the hub shell by the frictional contact of two sets of plates which are alternately connected to the clutch body and the hub shell, and are normally kept in contact by a set of springs.

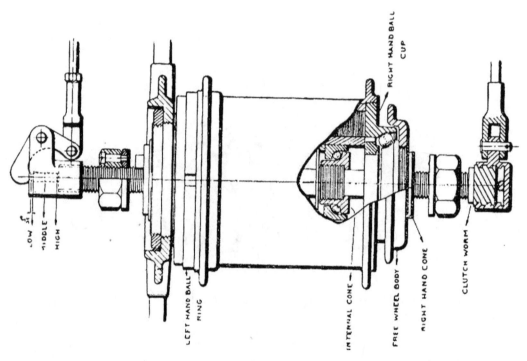

Fig. 131.—Collective View of the "Sturmey-Archer" Three-speed Gear.

When the spring pressure is released the gearing may revolve freely without driving the hub.

Is a Variable Speed Gear Necessary?

Variable speed gears naturally add some complication to machines equipped with them, but the parts are usually so substantially designed and well-made and assembled that little trouble is experienced with them. The presence of such mechanism on the motor cycle gives the rider a feeling of confidence when travelling over difficult or crowded roads, and conduces to better driving, greater all-round convenience, and much more pleasure than would otherwise be possible. At the same time, where expense is a serious object or the fear of added mechanism to look after acts as a strong deterrent, the prospective motor cyclist may be assured of

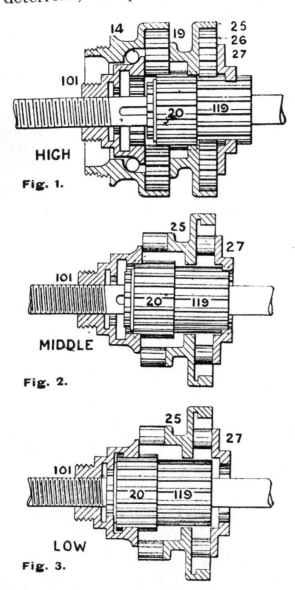

a large measure of satisfaction from the use of a single-geared machine —of course, fitted with an adjustable pulley. Where, however, the machine is to be used for side car work, and the frequent carrying of two riders, change speed mechanism of one sort or another may be regarded as indispensable, although a good driver with a free engine, single-geared motor cycle at his disposal can often obtain wondrous results with a passenger combination.

Variable speed mechanism is in growing demand among all classes of motor cyclists, and it often proves a difficult matter to dispose of one's machine in the second-hand market unless it is fitted with gearing of this description.

FIG. 132.—Internal Arrangement of the "Sturmey-Archer" Three-speed Gear.

As regards the use and manipulation of multiple gears, the the reader will find a few simple directions in Chapter XII, which deals in part with " Hints on the Driving of the Motor Cycle." Further and more detailed information on this and similar matters may also be gathered from a perusal of the companion volume entitled " How to Drive a Motor Cycle," obtainable from the same publishers.

CHAPTER IX.

LUBRICATION.

THE lubrication of a motor cycle engine is a matter requiring the most careful attention at *all* times; for not only does the efficiency of the engine depend very largely indeed upon it, but a distinct element of danger to the working parts arises should any serious neglect occur in this direction.

In the case of a steam engine the pistons, slide-valves, and other rapidly reciprocating parts work under much more favourable conditions than is possible with an internal combustion engine, for there the steam itself imparts a certain amount of lubrication and moisture, whereas in a petrol engine—in which everything works at a high speed and in a very high temperature—the tendency is to run dry unless the parts are continuously and adequately supplied with a proper means of lubrication. In the superheated locomotives now coming into favour on railways, special lubricants have to be employed for the cylinders and valves because of the dryness of the steam under high superheat; the steam at the highest temperatures partaking almost of the nature of a gas and containing many of the characteristics of petrol and similar vapours.

The means employed for lubricating motor cycle engines is known as the splash system, the oil being pumped into the crank-case, where, by capillary action, the flywheels, in revolving, throw it upwards and in various directions, so that it reaches every bearing and every part requiring to be lubricated. The engine thus, in a sense, looks after itself so long as a sufficient level of oil is maintained in the crank-case for distribution by the flywheels and the motor cyclist must always make it his particular business to see that such a level *is* maintained.

A portion of the tank of the motor bicycle is set apart for carrying a supply of lubricating oil, generally about one quart, but less in the case of lightweight machines with small engines. A hand force pump is provided, by means of which the oil is drawn from the lubricating compartment of the tank and forced into the crank-case through a pipe connecting the pump therewith. The oil compartment and pumps are nowadays placed, as a rule, near the after end of the tank so that the plunger may be conveniently operated by the rider, instead of, as formerly, being carried near the steering head, when it was necessary to bend forward with outstretched hand to reach the plunger. The precise construction can be gathered from an examination of the folding plate.

Character of the Lubricant.

A great deal depends upon the brand and consistency of the lubricating oil used, and readers are *specially* cautioned against accepting oil from a big drum bearing no name, and having nothing but the wayside vendor's recommendation to back it. The makers of the machine are always ready to suggest to the purchaser which particular brand and grade of lubricating oil is most suitable for the engine ; but this apart, it is quite safe to employ any of the well-known brands which can be purchased in distinctly coloured sealed cans bearing the manufacturer's name and trade-mark. Air-cooled engines require a much thicker oil than water-cooled ones, the reason being, of course, that the great heat generated within the cylinder is, in the former case, entirely dependent for its dissipation upon the surrounding currents of air, and the greater the heat the more rapidly will the oil be thinned and the limit of its lubricating properties reached. If too thick an oil be used the tendency of the piston to " gum up " in the cylinder when the engine has cooled down after working is increased, and the difficulty of starting again is largely augmented. On the other hand, if the oil be too thin the engine, although given a surplus supply, will not be adequately lubricated, and wherever a chance presents itself the oil will escape from the crank-case and splash all over the outside of the engine. And it should never be forgotten that it is the inside and not the outside of the engine that requires lubricating, so that all jointings must be kept tight and the tappets be a good (though freely working) fit in their guides to prevent oil splashing out through them. After much experimenting with various brands, the author has settled down to the use of the Vacuum Oil Co.'s "T.T." brand for the winter and the same makers' heavier grade for summer use.

The Effects of Over (and Under) Lubricating.

If too much oil be given, the carbonised deposits on piston and cylinder will be excessive, and the engine may at times be caused to misfire or stop altogether, owing to the sparking plug points getting fouled. A well-made plug will often spark quite efficiently in oil; but when over-lubrication is persistently indulged in, sooner or later (and generally sooner than later) a particle of burnt matter gets carried along with the splashings of oil to the plug and lodges between the points and the central electrode, thus short-circuiting the current by doing away with the air-gap and stopping the ignition of the mixture. It is immaterial when this happens, whether the sparking plug has one point or more, the short-circuiting being complete if only one of the gaps gets filled in as described. Twin engines, in which the sparking plugs are placed more or less horizontally, are specially prone to this trouble.

But the evils arising from *over* lubrication are as nothing compared with those which *under* lubrication may bring about. Not so much the constant sparing use of the oil, but real neglect of the lubrication for long periods at a time. Presumably, no one possessing the least common-sense would do such a thing wilfully, yet it remains a fact that splendid engines have been damaged, and some actually ruined, by inadequate use of the oil pump. If the engine becomes denuded of oil altogether, and is made to continue working at a high speed in that condition for any length of time, the natural consequence is that either the bearings or the piston (perhaps both) will seize up and become wholly immovable until drastic methods are employed to release them; and once any part has seized it is generally a matter for the repairer, while a new cylinder altogether, to say nothing of other parts affected, may have to be fitted and a considerable expense thereby incurred.

Channels or oil-ways are cut on the insides of the crankcase, and wherever it is necessary the oil should work its way—that is, between the shafts or pins and the bearings in which they work—and the splashing of the lubricant is so thoroughly performed that not only the main portions of the engine obtain a sufficient supply (when a proper level is maintained in the crank-case), but the timing and magneto gears are lubricated as well.

Automatic Lubrication.

The hand-pump method of lubrication is at best rather a crude one. It leaves a lot to be decided by the "human element," in the form of the driver's individual attention; and with a view in large measure to superseding this, the makers are in some instances fitting sight-feed drip lubricators, by means of which the lubrication of the engine is automatically performed, the only matter with which the driver need then concern himself being the primary one of setting the rate of feed, which is effected by means of a small needle valve, adjusted by a handle, as seen in the sketch, Fig. 133,

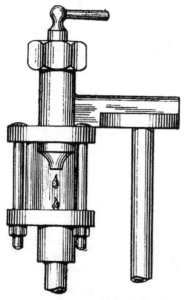

which illustrates a J.A.P. automatic lubricator, the functions of which are as follows: By means of a non-return valve a partial vacuum is set up in the oil-pipe and a steady and even supply of oil is drawn through a needle valve and glass barrel. The latter enables the rider to see at a glance if the oil is coming through, and in what quantity, and a small tap controlling the needle valve adjusts the supply. The lubricator, once set, is quite automatic, and the faster the engine runs the more oil is admitted, or the reverse. At the first outset, with a new machine, it is advisable to open the needle valve one-and-a-half turns. This will give plenty of lubrication and

Fig. 133.—" J.A.P." Automatic Sight-feed Lubricator.

it can afterwards be economised to the driver's own ideas.

In warm weather the drip system may be relied upon entirely, and it has the great advantage of maintaining a constant flow of oil to the engine, instead of there being, as already pointed out, a surplus part of the time and a shortage at other times. The rate of feed can be automatically increased when extra work is called for, and decreased when the conditions are normal or less arduous ; and only in very cold weather is there likely to be any difficulty in getting the device to work properly. This automatic system is usually supplemented by an ordinary hand pump, so that should any derangement arise, the older method can be fallen back upon, and in some cases the two systems are actually combined in one fitting.

Symptoms of Oil Shortage.

The motor cyclist should make it a practice, whilst riding, to keep his ears open for any unusual sounds arising from the engine, and if in any doubt whatever the best plan is to inject a charge of oil. Should a decided acceleration follow, it may be taken as certain that the engine was calling for more oil, and the unaccustomed sound will probably disappear almost at once. Preliminary signs of overheating are a decided slowing down and labouring of the engine, coupled with a more or less metallic noise, and should these symptoms develop, the remedy must be quickly applied, for otherwise the next thing to happen may be a partial or total seizure. In these circumstances—*i.e.*, when the indications are pronounced—it is well to stop and allow the engine to cool down and charges of oil should be worked into the bearings by pedalling before starting the engine up again.

The author once came up with a motor cyclist who complained, as he rode along, that his engine seemed to be falling off in speed. Asked how long it was since the last charge of oil had been injected, he said he had quite forgotten. On two successive charges being given, the engine at once accelerated in a most remarkable manner, and tackled a steep hill immediately afterwards in splendid style. These spurts, after lubricating, *always* indicate oil shortage, and it is the rider's business to make sure that the same thing does not recur.

General Hints on Lubrication.

Attention should occasionally be given to the force pump, the plunger packing requiring to be a good fit in the barrel so as to exert a full suction to draw a charge from the tank ; and where the pump is enclosed, as is nowadays often the case, the rider can only tell by the feel when the parts are working properly. A cock is provided at the base of the pump, the handle of which must be turned in one direction (usually horizontally) when drawing oil from the tank, and in another direction (usually downwards) when actuating the plunger for injecting the charge into the crank-case. The frequency of oiling the engine and kindred matters receive attention in a succeeding chapter. It is largely a matter for the individual, and he must use his common-sense as to when the supply should be increased or decreased. Much better over- than under-lubricate, and, best of all, do neither ; but strike a happy medium. Drain off the stale oil from the crank-case at the end of a long run, and always make sure that the oil-retaining screw in the latter is in place and properly screwed home. Should it fall out on the road and the fact be discovered, use anything that comes to hand, preferably a cork, to stop up the aperture until a properly fitting plug can be obtained ; otherwise the oil will work out through the opening on to the road and be lost for the purpose of lubricating the engine. In fitting a temporary plug, make sure that it clears the flywheels. Periodically the engine should be swilled out with paraffin, and after this has been completely drained off three or four full charges of fresh lubricant should be given.

CHAPTER X.

THE SPRINGING OF MOTOR CYCLES.

THOSE who have never ridden a motor bicycle are apt to allude in tones of disparagement to the "fearful" vibration to which users of that class of machine are, according to them, subjected, and until within the last few years there may have been some point in their criticisms. As in every other direction, however, the springing of the motor cycle has now been very materially improved; but, if there *is* any direction in which the majority of manufacturers have still some opportunity to afford us a wider maximum of comfort, it is in regard to the rear springing; for, although the saddles turned out by the leading firms manufacturing that class of goods are scientifically and beautifully made, it can hardly be said that the ideal has been so nearly approached, where the after part of the machine is concerned, as has been done in connection with spring front forks.

There are several different forms of the latter, most of them efficient, and some superlatively so. What is required in a spring fork is elasticity of movement in the vertical direction, without rolling or sideways action, and this fact is now fully appreciated, and provided for in the majority of cases. The front springing must be such as to insulate, so far as is reasonably to be expected, the rider's hands and arms from excessive vibration, even on the roughest roads, without, however, introducing a bouncing or rolling action, and thereby taking away from the ease and stability of the steering. Simplification of construction and moderate cost have also to be considered, but a few makers have not hesitated to provide a spring *frame* as well as spring forks, and, where this is done, the rider's comfort is appreciably added to. The author believes himself to be correct in stating that the spring movement should be in a line approximately with the steering angle, and the rearward movement should increase with the speed of the machine; but for general purposes it may be taken that a fork moves upwards and rearwards at an angle of about 65 to 70 degs.

Compression springs have this advantage over tension springs—that if one breaks it only closes a space equal to the distance between each coil—*i.e.*, about $\frac{1}{4}$ in. or 5/16 in. The position of the springs is of the greatest importance. They should be placed in such a position that the shocks are directly absorbed by the inertia of the weight of the machine. The position of the springs in some forks increases the strain on the latter and the column, instead of relieving it.

Fig. 134 illustrates the new spring handle-bar recently introduced by the Sphinx Manufacturing Co., of Birmingham,

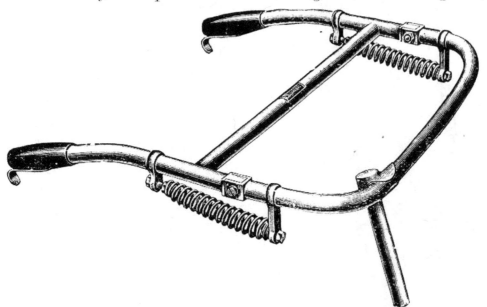

FIG. 134.—The "Sphinx" Spring Handlebar.

for use on motor cycles. In this device the springs are adjustable and the lateral bearings easily tightened. A total range of action of about 6 ins. is provided. When mounting, the bars are fully depressed and perfectly rigid, while in use the weight of the rider's arms is sufficient to poise the springs and insulate the arms from shock. The makers claim for this spring handle-bar a combination of lateral rigidity with vertical resilience, and they emphatically assert that it entirely prevents the ill-effects of vibration.

A thoroughly efficient type of spring fork is the "Druid," in which the parallel motion of the forks in relation to the steering column is controlled by means of links pivotally connected to lugs on the steering head, both at top and bottom, and to other lugs brazed on to the forks, these links moving in conjunction with a pair of coiled springs attached to the rear of the main fork. Thus are obtained two rectangles, placed, one at the crown, and the other at the top of the ball-head. The spindles forming the two long sides of the rectangle are firmly fixed in the links forming the two short sides, so that the links do not work upon the spindle ends, but the spindles themselves oscillate in their bearings. As the fork itself is carried on the front portion of these rectangular bearings, it is clear that, as soon as an effort is made to steer the machine, the response of the front wheel must be immediate. The springing effect is admirable, and an advantage of this type is that it can be applied—in varying proportions, of course—to practically any motor bicycle, irrespective of horse-power, weight, or other consideration. In some other designs the springs are arranged vertically, just in front of the steering head, and embracing a column forming a part of the fork, there being, in addition, links arranged for pivoted motion and connecting the forks with the steering head.

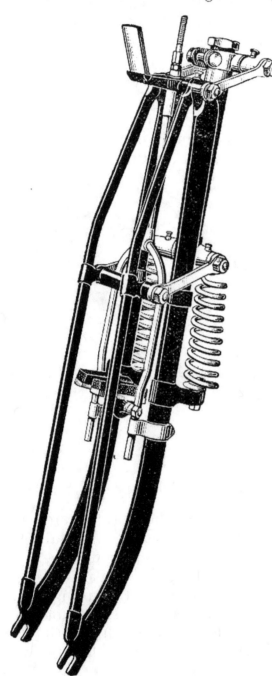

FIG. 135.—The " Druid " Spring Forks for Motor Cycles. A much-used type.

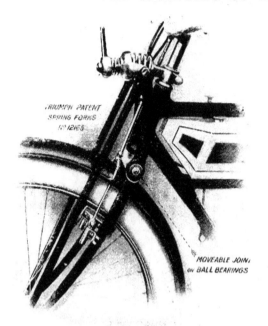

FIG. 136.—The " Triumph "
Spring Fork.

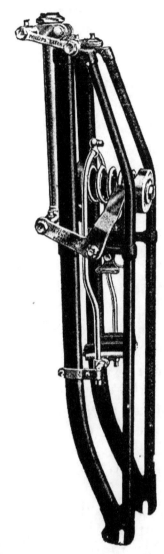

FIG. 137.—The " Saxon "
Spring Forks.

FIG. 138.—The " Indian " Laminated
Spring Fork.

Another example from modern practice is the " Triumph " spring fork, illustrated in Fig. 136. The forks in this design are pivoted on ball-bearings at the crown, which gives a perfectly free motion with only one movable joint, and maintains lateral rigidity (the importance of which has already been emphasised) of the forks.

The springs act in a fore and aft direction instead of vertically, as in the previous instances, and the wheel always remains in the same position. Arrangements for adjusting the tension on the springs are provided, and the correct tension is easily ascertained by riding fairly fast on a rough road to see whether vibration is effectively disposed of or not. These forks are fitted exclusively to " Triumph " machines.

Summarised, the points to be aimed at in designing spring front forks for motor cycles are as follows :—

(1) Absolute lateral rigidity and instant response to steering.

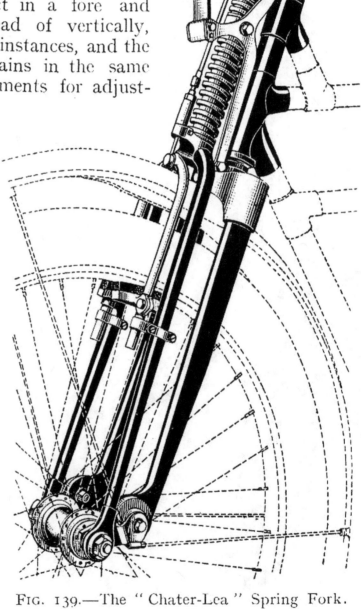

FIG. 139.—The " Chater-Lea " Spring Fork.

(2) The spring movement should be approximately in line with the steering.

(3) It should allow of the application of a front brake, either band, hub, or rim.

(4) The mudguard should move with the wheel.

FIG. 140.—Spring Fork fitted to "Scott" Two-stroke Motor Cycles.

(5) The breaking of a spring or springs should not affect the safety of the rider or prevent the machine being ridden.

13

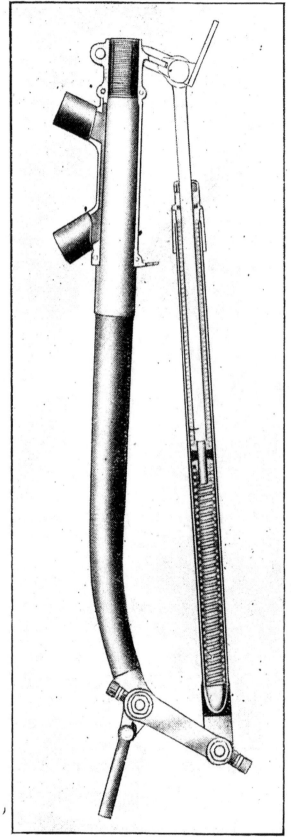

FIG. 141.—The "F.N." Spring Fork.

(6) The easy detachment of the wheel should be provided for.

(7) Shocks should be absorbed without a "galloping" action being set up after passing over an obstruction.

(8) The strain on the steering column should be relieved.

Spring Frames.

On the "Bat" motor cycles, and more recently the N.S.U. machines, the frame itself participates in the system of springing. Each of these arrangements is illustrated. In the first instance, the foremost members of the forks are made of tubular section. As the wheel moves upwards or downwards, these tubular members work on plunger rods, connected with the head, in conjunction with coil springs, as seen in the drawings. The same arrangement is employed at the rear of the frame, and the chain stays are carried at their forward ends on a cross ball bearing spindle to permit of the springing action taking place. Here also the piston-rod method is combined with coiled springs in compression,

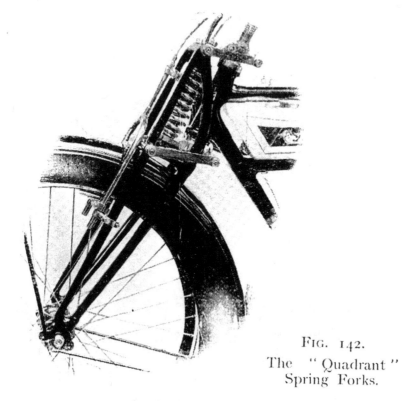

Fig. 142.
The "Quadrant"
Spring Forks.

Fig. 143.—The Luxurious Front Springing of the "T.M.C."
Four-cylinder Machine.

and the result is to largely increase the ease and comfort of the rider.

The "Bat" spring frame is illustrated in Fig. 144. The seat column is free to move vertically and independently of the frame. It is held firmly to the latter by four links which are mounted upon ball-bearings. It is then held in position by four spiral springs attached at one end to the frame and at the other to the two top links. Footboards

Fig. 143A.—Rear Springing of the "T.M.C." Four-cylinder Motor Cycle.

are mounted on the two bottom links. As the rider mounts the springs gently yield to the varying pressure which his weight produces. The front forks are similarly contrived to isolate the front wheel from the rigid portion of the frame, and the effects produced by unevenness in the road are absorbed by four spiral springs. The front fork ends are provided with adjustable ball-bearings on which an arm carrying the necessary sliding spring holders is mounted

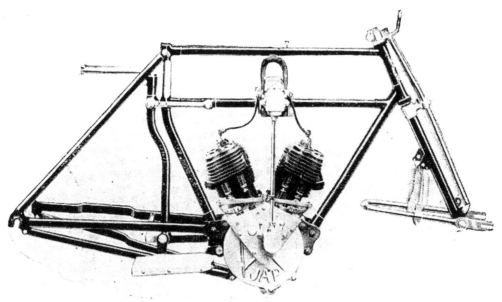

FIG. 144.—The "Bat" Spring Frame.

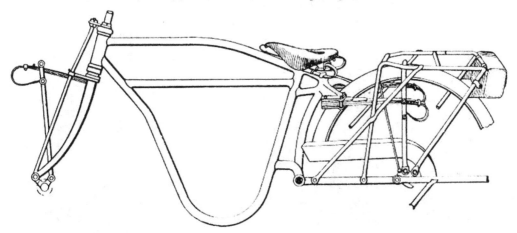

FIG. 145.—Spring Frame of the "Indian" Motor Bicycle.

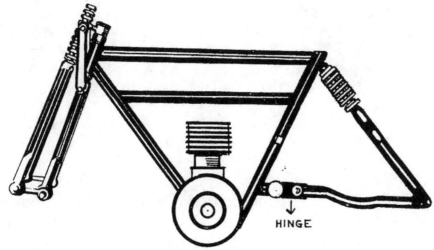

FIG. 146.—Diagram of the "N.S.U." Spring Frame.

and slotted to take the front wheel, and passes entirely round the wheel at the back. The wheel is held rigidly in the horizontal position, but allows free vertical movement ; the latter being held in check by the four springs. These springs can be adjusted to any conditions by means of the

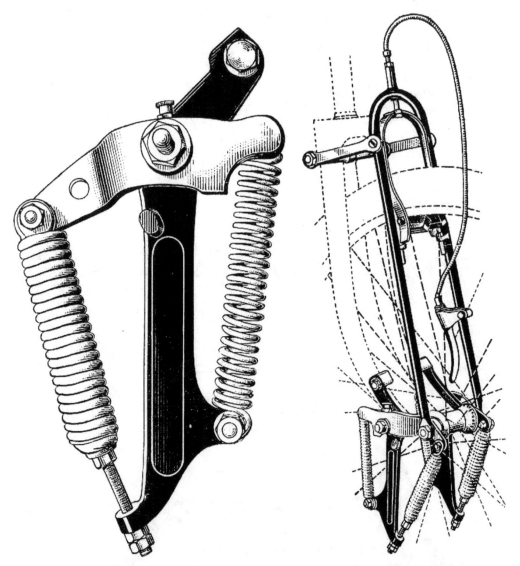

FIG. 147.   FIG. 148.
The " XL-All " adaptable Springing Device for Motor Cycle Front Forks.

sliding holders on the arms. In the latest " Bat " machines, the number of front fork springs is reduced to two of larger size. The " N.S.U." spring frame has the chain stays hinged just behind the bottom bracket, and there is a telescopic connection between the bridge of the upper stays and the

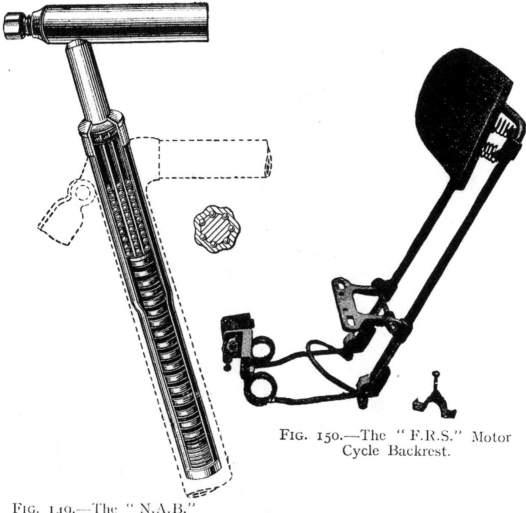

FIG. 149.—The "N.A.B." Spring Seat Pillar.

FIG. 150.—The "F.R.S." Motor Cycle Backrest.

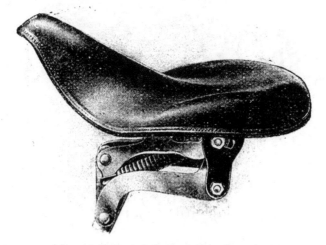

FIG. 150A.—The "XL-All" Saddle, Cantilever Springing.

saddle pillar lug.   A strong compression spring surrounds
the telescopic member.

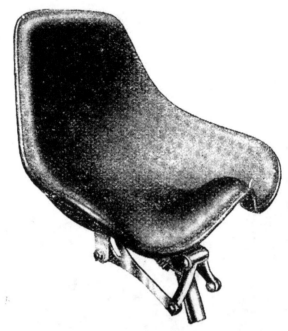

FIG. 150B.—The " XL-All " Pan Seat Saddle.

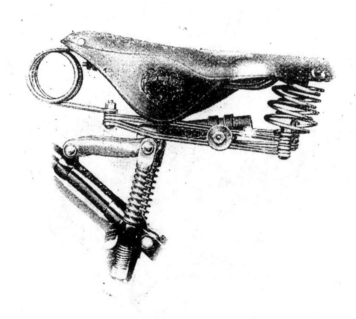

FIG. 151,—The " Rex " Cantilever Spring Saddle Support.

Adaptable Springing Devices.

In addition to the springing devices sold integrally with
the motor bicycle, there are others supplied by firms of

accessory dealers and manufacturers for adaptation to any machine on which rigid forks or other similar out-of-date fittings are employed ; as, for instance, old pattern machines which the riders desire to bring up to date. These include spring attachments for the front forks, spring seat pillars, and saddles. The " XL-All " specialities of this description are often seen in use, and some of these are included in the appended illustrations.

The Rex Company also provide their touring motor

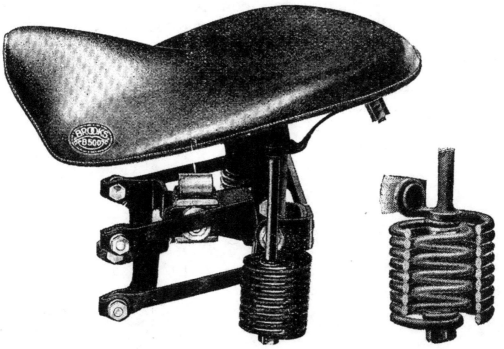

FIG. 152.—The Latest Type of Brooks' Motor Cycle Saddle with Compound Springs.

FIG. 153.—Compound Springs of Brooks' Saddle.

bicycles with a cantilever spring saddle support, and a reproduction of this is also given on page 200. The A.S.L. firm fits motor bicycles with an air-springing arrangement, comprising a piston moving freely within a cylinder, into which air is introduced to give the necessary resistance and elasticity. This system of springing is adapted for use both at the front and rear portion of the machine. Purchasers of motor cycles should see that the springing arrangements are adequate. As regards saddles, it would be very hard indeed to beat those of Messrs. J. B. Brooks, Ltd., of Birmingham, fitted, as shown in Figs. 152-3, with compound coiled springs and padded tops. These saddles are produced with different strengths of spring to accommodate different riders' weight, and the whole of the construction is of the highest class. The XL-All saddles (Figs. 150A and 150B), constructed on the cantilever principle, provide an extreme degree of comfort. They are excellently made and can be strongly recommended.

The F.R.S. Automatic Backrest for motor cycle saddles (Fig. 150) adds much to the comfort of the rider, especially when a long journey is undertaken. It falls down on to the carrier of the machine the moment the rider lifts his weight off the saddle to dismount, and is brought into action by pulling it up to one's back after getting seated on the machine. After this it is perfectly automatic. With this appliance in use the rider is enabled to make the longest journeys without experiencing the slightest inconvenience from back-ache.

CHAPTER XI.

OVERHAULING AND TUNING UP THE ENGINE, ETC.

WHEN a new motor cycle is delivered to its owner the engine is, or should be, in the pink of condition, and every endeavour should be made to keep it so. Of course, as wear takes place, adjustments become necessary; but it cannot be too repeatedly or forcibly urged that so long as everything is working properly the best plan is to avoid interference with

FIG. 154.—Overhauling is both interesting and instructive.

the maker's adjustments, though it amounts almost to a duty to go carefully over the machine to see that all nuts are properly tightened up.

In the case of a second-hand machine, it may be advisable to overhaul the engine and other parts, with a view to ascertaining their exact condition; but even this may very well be deferred so long as the machine runs well and does what is required of it. It is the author's purpose in the present chapter to deal briefly with the management of the machine, both at home and when out on the road ; to briefly consider the more important adjustments, and how to make them ; and generally to treat of the subject as broadly as is possible in the space at command, from the point of view of those whose knowledge of motor cycle management is either limited or altogether wanting. Some slight reiteration of previously stated points is necessary in places, but this, it is hoped, will be excused in view of the extreme import- ance of making this portion of our subject especially clear.

Diagnosing Faults.

If the engine is suffering from some definite complaint— such as loss of compression, faulty carburation, or ignition trouble—the symptoms are usually pronounced, and it is possible, as a rule, to rectify the trouble by devoting one's attention entirely to the affected part ; but, where general debility is the mischief, a system of "tuning up" all round has to be undertaken, and this may involve the overhauling of several vital organs in turn, each one of which has become lacking in its ability to properly perform its allotted function. And then, again, it is not necessary to wait for the develop- ment of a fault before undertaking an important adjustment. The careful motor cyclist always takes a good look round his machine before setting out on a ride of any length, and in this way he often detects—and puts right in the shed before starting—something which, had it been neglected until out on the road, might have caused delay and bother, and which, in any case, could be more conveniently dealt with at home.

The Maintenance of Power.

The makers design their engines to develop a certain horse-power at a given speed in revolutions per minute, and anything below that speed means a lowering of the horse- power at disposal. Therefore, it must always be the endea- vour of the owner to maintain the ability of his engine to reach the necessary speed to give its maximum horse-power when required, and unless all vital parts are working effi- ciently, it is impossible to secure this happy result. A sound frame, comfortable saddle, and efficient tyres, are all highly necessary to the success of motor cycling ; but given these, and at the same time a poorly conditioned engine, but little real pleasure can be obtained from the use of the machine, and the whole thing becomes a source of anxiety and trouble and, maybe, expense into the bargain. How necessary, therefore, that every motor cyclist should think and act for himself in the maintenance of his machine, and not have

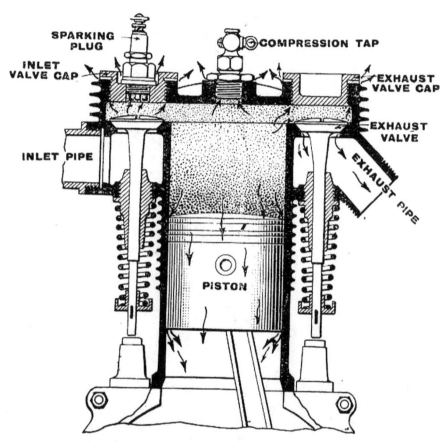

FIG. 155.—The " Compression " escapes wherever it can find the smallest outlet.

to go running to the repairer for every little adjustment and for the rectification of every little fault as it arises.

The Fit of Piston Rings.

To begin with, there cannot be speed and power unless the *compression* is good. Therefore, see that the piston rings are a good springy fit in their grooves, without vertical play therein, but able to expand and form a gastight contact with the walls of the cylinder. If the petrol gas is enabled to escape past the piston rings to any appreciable extent, the fact will be denoted by the appearance on the rings of discolorations resembling burns, and this should be taken as a sure index of loss of compression. The remedy is to replace the ring or rings with other and better fitting ones, and the opportunity should be embraced while the grooves are free to clean them, as if the engine has been working for some time without attention, accumulations of carbonised oil will be found adhering to them, and this interferes with

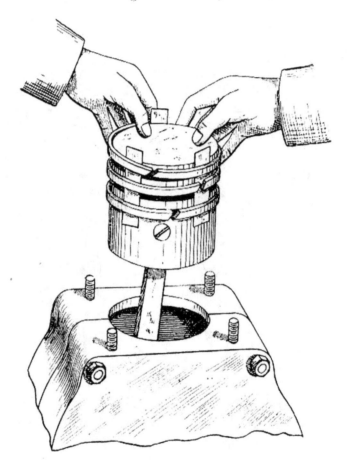

FIG. 156.—How to remove Piston Rings without breaking them.

the fit of the rings. To remove the piston rings it is a good plan to cut three pieces of tin, about 1 in. wide, and insert these at regular distances around the piston, and between it and the rings. The latter can then be worked off gently by the fingers without fear of breakage, and the same plan can be resorted to when replacing the rings or fitting new ones, although, generally speaking, it is easier and less risky to fit rings on than to take them off. Until the new piston rings have " worked themselves in," the compression will in all probability remain faulty. A hundred-mile run will, however, put matters right in this direction.

Piston rings are made of cast-iron, and are, therefore, easily broken. They cost from 1/- to as much as 1/9 each ; so that it is just as well to be careful in handling them. Some engines have " stepped " rings (see Fig. 36), the slots being undercut, and, in some designs, a small pin is inserted in the groove so that rotation of the ring in its groove is prevented. Care is necessary with the ordinary type of ring with diagonally cut slots to see that the slots do not lie in

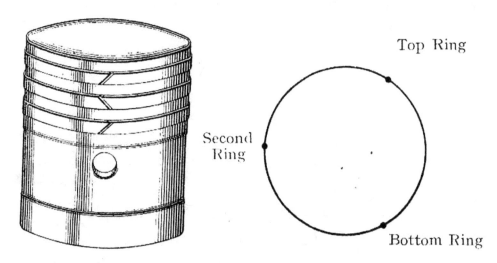

Top Ring

Second Ring

Bottom Ring

FIG. 157.—The wrong way of setting the Piston Ring Slots.   The right way.

the same vertical plane when replacing the cylinder, for if this occurs there is likely to be some loss of compression ; but all the care in the world will not prevent their sometimes *working* into line of their own accord while the engine is running, if they are at all a slack fit. With the stepped and pinned type of piston ring, illustrated in a previous chapter, this cannot, of course, occur. After a very large amount of use has been made of the engine—say, at the end of some 40,000 miles or more—it may become advisable to re-bore the cylinder and fit a new piston and rings. There is a tendency for the cylinder to lose its parallel bore after long usage, and when this has occurred nothing will ensure a good compression except the aforesaid measures, and this, of course, is a job for a competent mechanic alone.

Grinding in the Valves.

Having made sure that the piston rings are both well fitting and properly aligned, the next point towards maintaining a good compression is to keep the inlet and exhaust valves properly ground into their seats so that leakage cannot take place there either. The exhaust valve is naturally the chief offender in this respect, for it has to withstand the constant passage, between its surface and the seating, of burnt gases at a very high temperature and velocity, whereas the inlet valve only has clean, unburnt gas to deal with, which, as a matter of fact, imparts a certain amount of lubricating influence to the valve. The action of the

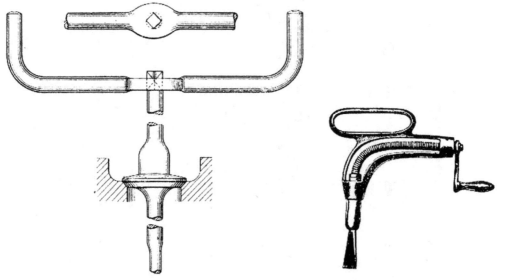

FIG. 158.—Special Tool for grinding   FIG. 159.—The " Service "
Motor Cycle Engine Valves.   Valve Grinding Tool

exhaust gases is to produce, in course of time, abrasions and pit-markings on the exhaust valve and its seating, and when this has occurred to any marked degree the ability of the valve and seating to prevent leakage of the gas past them during the compression stroke is impaired and may become seriously so if the grinding operation is not undertaken.

It is a very easy matter to grind in an ordinary mechanical valve—all that is necessary being to remove the spring and cotter, smear a mixture of oil and emery, or other prepared powder, such as carborundum, on its surface, and then rotate the valve in its seating, either by means of a screwdriver or a brace, until both the surface of the valve and

the seating which receives it are freed from pit-marks and
both present a smooth, even surface all round. At the same
time, this is just one of those matters which require doing
properly to get the best results. The valve must, during the
grinding operation, be lifted at intervals and allowed to
assume a different position in the seating, and the grinding
medium must be renewed from time to time as the work
proceeds.

It is a good plan to place a weak spring—as, for instance,
an old A.O.I. valve spring—under the head of the valve,
so that directly the pressure on the screwdriver or other
instrument is relaxed, the valve will rise automatically, and
can be either lifted out or rotated slightly, as the operator
desires. Many motor cyclists make special grinding tools for

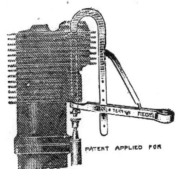

FIG. 160.—Valve Spring Lifter.   Useful when
removing or replacing a Mechanical Valve.

FIG. 161.
Terry's Patent Valve
Lifter.

themselves, and the author has found those illustrated in Figs.
158 and 159 very useful for the purpose. Continuous grinding
in one position often produces ridges on valve and seating,
and may quite possibly make matters worse than they were
at first. It is sometimes difficult to get the spring and
cotter free when setting out to do valve grinding. Special
tools are sold for this purpose (Figs. 160 and 161), and the few
shillings they cost probably repay the purchaser in the long
run, if only by saving the cost of sticking plaster for barked
knuckles. Always place a piece of rag in the port opening
while grinding in a valve, in order to prevent the smallest

14

particle of the abrasive grinding mixture from entering the cylinder.

The "Lift" of M.O. Valves.

The next point—and a *very* important one—is the maintenance of a proper gap or space between the bottom of the

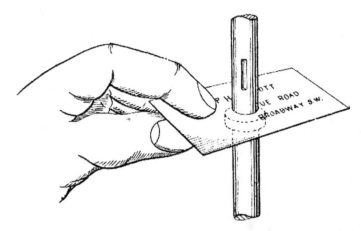

FIG. 162.—There should be just sufficient space for a visiting card to pass between the Valve Stem and Tappet with engine cold.

valve stem and the face of the tappet which operates it. In course of time the valve stem wears and shortens, and the gap is then increased, so that the valve does not receive its full opening. Mechanical valves require about ¼-in. lift off

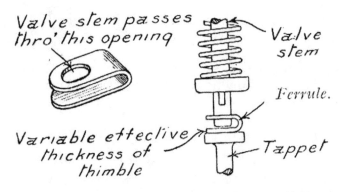

FIG. 163.—The "Service" Tappet Adjuster. An Excellent Device.

their seats, that representing, in the case of an average valve, about one quarter of the effective diameter. The timing-gear cam and other parts are designed to provide this amount of opening, but the tendency is, of course, for it to become less as wear takes place. The space between valve stem and tappet should be just sufficient to allow a visiting card of ordinary thickness to be passed, not *forced*, through with engine cold ; and this being so, when the valve heats up and the stem slightly expands, the gap will be, to a very small extent, reduced and the full valve lift assured.

When the valve stem has worn and the space is consequently increased, matters may be rectified either by brazing an extension-piece on the stem and then filing down to the exact length required, or other means of effecting the same result may be employed. The ferrules, or, as they are called by the vendors, " tappet adjusters " (shown in Fig. 163), are extremely handy for effecting this purpose. When the tappets themselves have become markedly worn they should be replaced by others, and here the cost is a very moderate one. The cams

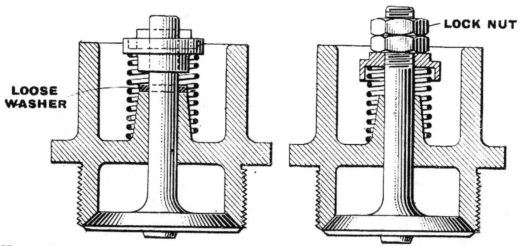

FIG. 164.—Two methods of securing and adjusting A.O.I. Valves. (See accompanying description.)

only require renewing at long intervals, a huge mileage being usually put up before any such measure becomes necessary as the result of the wear. A new rocker will, to some extent, make up for wear on the cam, and is cheaper. A few makers fit their engines with adjustable tappets, and, provided the locking arrangements on these are well designed, the idea has much to commend it. If not, trouble may ensue, as the result of the tappet automatically adjusting itself when not required.

A new valve will generally be found to possess too long a stem, and it is highly necessary that this should be cut to *exactly* the right length, for it is easy to see that if the length be only ever so slightly in excess of what is correct, the stem and tappet will be brought permanently in contact and the valve will be prevented from bedding into its seat. A definite loss of compression will be the result, and it may be even impossible to start the engine at all. If, on the other hand, the stem is shortened too much, the evil previously described will arise.

The exhaust valve lifting mechanism sometimes provides a means of difficulty, especially for the embryo motor cyclist. Care must be taken to see that the lifting lever (or whatever form of device is employed for raising the valve) lies clear of the tappet at all other times than when deliberately brought into contact therewith by the rider at starting or otherwise in the course of manipulating the engine.

Valve Springs.

Loss of engine speed may arise from using springs of insufficient strength on mechanically operated valves. The beginner should never experiment in this direction, but should consult the makers of the engine if he has any reason to believe that inefficiency is arising from this cause. Too *strong* a spring, on the other hand, may result in valve breakage by bringing the valve down with undue force on its seating. It is very necessary that valves should be smartly and promptly closed at the proper times, but sledge-hammer methods are certainly *not* wanted.

Adjusting A.O.I. Valves.

The adjustments necessary in the case of an automatic valve are, perhaps, more easily carried into effect than where the valve is of the M.O. type ; but, at the same time, they require more delicate handling, as only a very slight variation makes a tremendous difference in the running of the engine, especially a twin-cylinder one. Some A.O.I. valves are secured, with their springs, by means of the cup and cotter method, much the same as in a mechanical valve ; but, in others, the stem of the valve is screw-threaded and lock nuts are employed for varying the tension on the spring. Both methods are illustrated in Fig. 164. The valve should be given a lift of about 3/32 in. at the start, and this, in course of working, will automatically increase to $\frac{1}{8}$ in., and more, if left for long periods without attention. A simple method of reducing the opening of the valve where the cotter method of fixing is employed is to place a washer, as shown in the drawing, on the column surrounding the stem, and with which the collar comes into contact as the valve opens, thus limiting the travel of the valve. With the lock-nut method, as shown in the second sketch, adjustments are easily effected by screwing the nuts in one direction or the other to give more or less valve opening as may be desired. As was pointed out in a previous chapter, a small automatic valve opening increases the speed of the engine, while a larger opening reduces it, but makes it possible to run the engine, firing regularly, at low speeds. If the opening be really excessive, what is known as "blowing-back" in the carburettor may occur. The inlet valves at such times admit more gas than the engine can take, and a portion of the

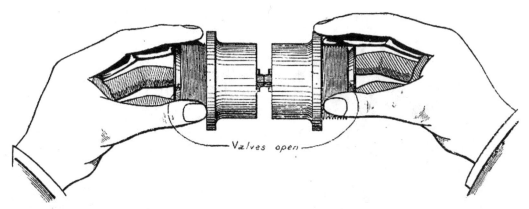

FIG. 165.—Testing Automatic Inlet Valves for opening.

charge is forced to return to the carburettor. This prevents the supply of suitable mixture being maintained, interfering with the action of the carburettor and the proper working of the engine. Automatic inlet valves practically never require grinding in, and when they do, the task is the easiest one imaginable, owing to the fact that the seating comes away from the engine with the valve. It is very necessary in a twin-cylinder engine that both A.O.I. valves should have precisely the same amount of opening and springs of identical strength, and a good way to secure this is to press the stems of the two valves against one another (each valve being complete with spring and seating), and the valve which opens first shows itself to have a weaker spring than the other, and must, if the other valve is getting a correct opening, be provided with a rather stronger spring, so that when pressing against its mate both open simultaneously with identical adjustments.

Cylinder, etc., Jointings.

All jointings—such as where the valve caps, sparking plug, and compression tap screw into the cylinder—must be made compression tight by interposing copper and asbestos washers, and to test these engine oil may be smeared around the joints, and the engine pedalled or otherwise turned over compression, when, if there is any leakage, it will show itself by the appearance of tiny bubbles caused by the escape through the oil film of the air or gas under compression. The task of removing the sparking plug, valve caps, etc., is sometimes fraught with difficulty, owing to the parts having become so tightly fixed in their places by reason of the heat they are subjected to that they cannot be loosened. Use a properly fitting " fixed " spanner and, if very obstinate, the parts should be well soaked with paraffin and then left for a while before again attempting to loosen them. A sharp blow on the end of the spanner with a hammer may be necessary.

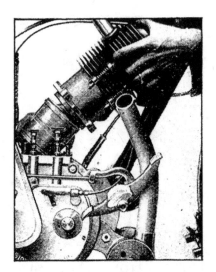

FIG. 166.—Removing the Cylinder of a New Hudson Motor Cycle Engine.

Carbonised Deposits.

Before leaving that part of our subject which deals with the care of the piston, cylinder, valves, etc., a few words may be added as to the necessity of removing the carbonised deposits which will be found both on the inside and outside of the piston head, the combustion chamber of the cylinder and other parts, when the cylinder has been removed. A clearance of these deposits is recommended at the end of every 1,500 miles (the author himself, who believes in rather overdoing than underdoing the lubricating of the engine, takes the job in hand, on the average, every 1,000 miles), and believes it to be beneficial and wholly worth while. If heavy accumulations of burnt oil, road dust, and whatever

FIG. 167.—Making a Cylinder Jointing by the " Template " method.

FIG. 168.—Dismantling the Engine of a " P. & M." Motor Cycle.

the other constituents may be, are left on the parts, a tendency to pre-ignition is induced, owing to the gases being fired by incandescent effect before the electric spark is timed to take place, and back-firing—*i.e.*, explosions, occurring before the piston is in a suitable position—results. The engine tends also to overheat, and, in any case, it is not working

FIG. 168A.—The "New Hudson" Decompressor Device Assembled. (See page 221.)

FIG. 168B.—The "New Hudson" Decompressor Parts.

under anything like such healthy conditions as when the parts are clean. The accumulations are easily chipped away by the aid of a screwdriver or other blunt instrument, the whole job occupying but ten minutes or a quarter of an hour at most, after access has been gained to the parts.

When clearing the top of the piston, it is well to stuff a piece of rag into the crank-case opening so that the burnt matter cannot fall through the latter and get mixed up with the lubricating oil when the engine is subsequently working. Some motor cyclists smear the top of the piston with graphite, with the idea of preventing, in some measure, the collection of carbon deposits, but it is doubtful whether they gain much by so doing.

Removing and Replacing the Cylinder.

Before removing the cylinder of a motor cycle engine, it is a good plan to squirt a few drops of paraffin or petrol through the compression tap. This will free the piston and simplify the task of getting the cylinder off. Take the cylinder in both hands, and, having previously moved the piston to the bottom of the stroke by means of the pulley, partly rotate the cylinder backwards and forwards on the piston, while at the same time pulling upwards. Different engines require different treatment and, in some, the best way to get the cylinder clear is to place the connecting-rod and piston at an acute angle fore and aft of the crank-case, so that more space is available than when the latter parts are vertical and the cylinder has to be lifted up directly underneath the tank.

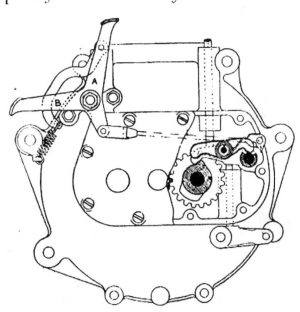

FIG. 168c.—Construction of the "Ariel" Decompressor.

When replacing the cylinder, make sure that the jointing material between it and the crank-case is in good condition; if not, it must be replaced by a fresh one. Before doing this, however, make sure that the two surfaces are clean and smooth. An efficient jointing can be made of brown paper soaked in oil, and, in preparing this, first cut a disc out to allow of the paper sheet being passed over the piston, and then shape out the jointing (and the four holes for the holding-down screws to pass through) by means of light tapping with a hammer. The work of cutting out a jointing with scissors is tedious and unnecessary. Remember, that if the thickness of the jointing is materially increased, it will tend to raise the cylinder and may interfere, if carried too far, with the gap between the tappet and valve stems. In

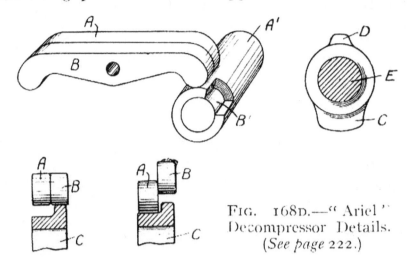

Fig. 168D.—"Ariel" Decompressor Details. (*See page* 222.)

some engines it is impracticable to remove the cylinder alone; the whole engine has to be taken down.

Tuning up the Carburettor.

Carburettor adjustments are necessarily rather delicate ones and need careful consideration at times. There is, however, no part of the mechanism associated with a motor cycle engine which should be spared the attentions of the inexperienced (sometimes rudely described as "tinkering") so much as this one. If the level of the petrol is practically flush with the top of the jet, with the machine resting on both wheels, and the carburettor fixed in its running position, and the needle valve forms a good joint in its seating, the proper sizes of jet and choke tube (or adapter) are fitted, and intelligent use is made of the extra air lever, matters will be found to be about right. Guidance in the direction of selecting the proper sizes of jet and adapter for particular engine

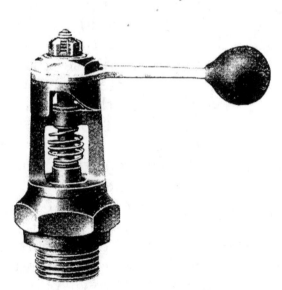

Fig. 168E.—The "Endrick" Decompressor. (See page 222.)

powers was given in the Chapter on " The Carburettor," and it is an easy matter with modern carburettors to experiment one size larger or one smaller if the engine shows a desire either for more or less air when running under normal conditions.

The motor cyclist must be guided by the symptoms displayed by his engine, and not rush, as the majority do, to the immediate assumption that the carburettor is responsible directly there is any falling off of power. Any interruption in the supply of fuel to the engine will naturally bring about erratic working, and perhaps a total stoppage ; but this may arise from other causes, among them being ignition failures and other temporary derangements.

Needle Valve Adjustments.

If the needle valve is a poor fit in its seat, overflowing of the petrol may result. Sometimes a little piece of grit gets

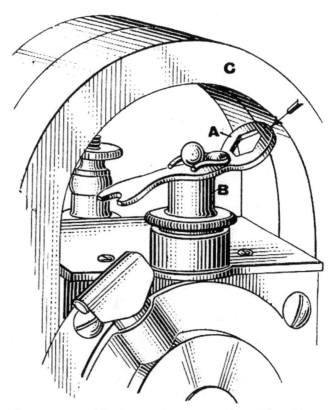

FIG. 169.—Testing an Accumulator by the lamp method.

FIG. 170.—Testing the current at the Magneto. (A) Spanner, (B) Carbon Holder and High-tension Terminal, (C) Magnets. Arrow indicates direction of spark across air-gap.

lodged in the seating under the valve and, this removed, matters assume their wonted condition. Then small particles of matter get lodged in the jet and partially stop up the bore, thus impeding or altogether stopping the flow of petrol. The rider should first ascertain that neither of these things has happened before he undertakes more drastic remedies. A few motor bicycles are fitted with petrol strainers designed to prevent the ingress of any particle of foreign matter in the petrol to the carburettor. The level of petrol in the jet can be adjusted by means of the brass collar on the needle valve ; if the position of this collar is lowered on the needle it effects the *raising* of the petrol level, but if it be lifted higher up on the needle the level will *drop*. Overflowing occurs when the petrol level is too high, and starving of the engine when it is too low. Underneath the cover of the carburettor float chamber will be found the two balance weights previously mentioned. Removal of the split pins holding these will liberate the needle carrying the collar (the latter being driven tightly in place but otherwise permitting of movement), and it is an easy matter to shift the position of the collar by lightly tapping it with a hammer, the needle itself being

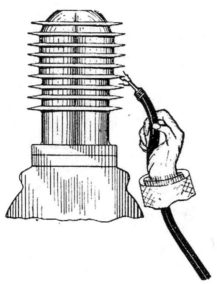

FIG. 171.—Testing the High-tension Cable.

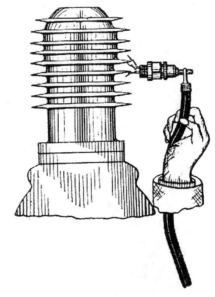

FIG. 172.—Testing the Sparking Plug.

preferably held in a vice. A very small adjustment makes a lot of difference, and the needle should be marked before anything is done, as an indication of what the original position was.

Difficulty in Starting.

If difficulty in starting the engine is experienced, and the ignition and other parts are known to be in perfectly good order, with the piston working freely in the cylinder, it may be that too much air is being admitted to the spraying chamber of the carburettor, thus weakening the mixture below explosion point at slow speeds when the magneto spark is weakest, and as a test a smaller adapter should be fitted. Raising of the petrol level (if below top of jet) may effect an improvement, or a larger jet may be needed. On the other hand, it may be wholly due to faulty manipulation of the levers, and experiments with varying positions of these will generally show which are the best to secure good behaviour at starting. Some engines require a fairly full throttle opening and the extra air shut right off ; others will not start unless extra air is admitted, while others again require some different setting. Usually throttle one-third open and extra air closed is about right.

In cases of partial or total stoppage, the first thing to do, before blaming the carburettor itself, is to make sure that there *is* petrol in the tank, and that it is getting through without impediment to the carburettor. Disconnect the petrol pipe union from the base of the float chamber and turn on the tap near tank, and if the flow is not a full one, remove the pipe altogether and seek for the cause. Also make sure that the garage man did not give you water instead of petrol (this happened once to the author) by allowing a small quantity to flow on to the ground or other flat surface, and if it quickly evaporates all is right. Always use a funnel fitted with a fine wire gauze strainer that will retain water but allow petrol to pass when filling the tank up. A piece of muslin will do as a makeshift. Ascertain if any leakage is going on between the tank and the carburettor, at the petrol unions or around the junction of the pipe with the tank, and endeavour to trace the cause of the trouble from its source onward to its discovery. If the extra air orifice of the carburettor is fitted with a gauze covering make sure that this is kept clean. If choked with dust the air cannot get through, and faulty running may result.

Certain makers equip their machines with what is termed a " decompressor "—a device having as its object the release of a portion of the gas from the cylinder during the compression stroke, thereby facilitating the process of starting the engine. The " New Hudson " decompressor is a good example of this practice (see Fig. 168A). On the outer face of the exhaust cam wheel (which is of the ordinary internal cam type) is mounted an auxiliary cam (A) sliding radially and working in conjunction with the exhaust tappet rocker. The said auxiliary sliding cam can be put in or out of action by means of an eccentric bush (B) protruding through the outside of the timing gear cover. On this eccentric bush is mounted a lever which in its turn is actuated by the ordinary Bowden wire mechanism from the handle-bar on the lines of, and mounted together with, the carburettor control.

The eccentric bush on being turned only slightly will reduce the compression in the cylinder by causing the exhaust valve to lift to a very large extent during the compression stroke, and if moved further will reduce the compression still more, so that an infinitely variable compression can be obtained.

The " Ariel " (Figs. 168c and 168d) is another decompressor incorporated in the timing gear. The rocking lever is formed in two portions lengthwise, the two halves (A) and (B) being connected or pivoted at the centre so that they may rise and fall together in one plane or separately in different planes. On the exhaust cam, and opposite to the main cam, is a nose-piece (D), and the rocking levers at the other end rest on a bush having a notch in it as shown in the sketch. If one of the halves of the lever rests on the solid portion of this bush, while the other falls into the notch, the effect is to lift one half higher than the other, while, when both rest on the solid bush surface, the two halves rise and fall together. The bush is rotated by means of a bell crank lever operated by the rider's foot, and when the two halves are in the same plane the nose at the back of the exhaust cam raises the rocking lever as a whole and releases the compression, while on rotating the bush, by pressing down foot lever (B), Fig. 168c, the end of one half of the rocking lever falls into the slot as above explained, and the engine works normally under full compression.

Yet another type of decompressor is that which screws into the valve cap of the cylinder as a separate fitting similarly to a sparking plug. This type can be adapted to almost any existing engine and is claimed to give excellent results (see Fig. 168e).

Ignition Faults.

Faults connected with the ignition (or, perhaps, it would be fairer to say : their predisposing causes) are numerous. They may be due to failure on the part of the accumulator or magneto to produce a sufficiently intense current, or the defect may arise somewhere between the generating medium and the point at which the electricity performs its function of igniting the charge. The high-tension cable or its connections may require attention or renewal, the sparking plug be amiss, or the contact-breaker have developed a fault which is preventing the engine from working. The usual *modus operandi* is to first test the accumulator, coil, or magneto to discover whether the current is being generated properly. The accumulator should show 4·5 by voltmeter, or, if the lamp test be applied, a bright light should

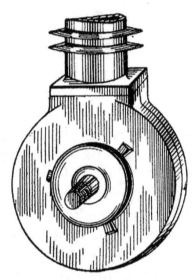

FIG. 173.—Wedges between Pulley and Crank-case assist in getting Pulley loosened for removal.

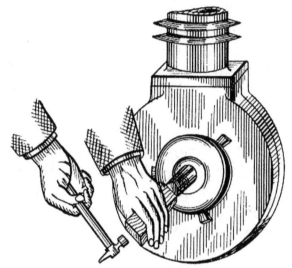

FIG. 174.—The Pulley should by this means come away quite easily.

be obtained, and the magneto, when tested by means of the small appliance provided for the purpose, should produce a bright electric spark. The appliance referred to is a small spanner with spring opening at the centre, which can be "snapped" on to the high-tension terminal in such a way that its two ends are each close to the inside of the magneto, as seen in Fig. 170. With the spanner in this position the armature should be rotated by using the pedalling gear or wheeling the machine along the road, the timing lever being set at full advance beforehand.

If the magneto itself is in proper working condition the spark will occur, and the fault must be sought either in the cable or sparking plug, and, as a matter of fact, the sparking plug is usually the first thing examined. If, however, no spark appears, the contact-breaker must be inspected to see whether the platinum points are properly making and breaking contact. When the fibre block passes out of the recess, the distance between the platinum points should be 0·5 mm. Should this not be so, the points must be adjusted by means of the screws provided for the purpose. Should, however, the amount of " break " be correct, the best plan is to remove the contact-breaker disc altogether by unscrewing the retaining screw, and if the platinum points are found to be dirty or pitted, they must be touched up with a smooth file or fine emery cloth until quite clean, making an even contact. The contact-breaker and the parts connected with it should always be kept quite clean (petrol is useful here) as nothing tends to interfere with their working more than an accumulation of oil and dirt. In the case of the new pattern single-cylinder Bosch magnetos, in which, as before described, the snap on terminal has been discarded in favour of the water-tight carbon-holder illustrated on page 106, the method of testing is as follows :—

The conducting wire should first be detached from the magneto and a fresh wire put into the carbon-holder and brought into such a position as to leave a distance of 1 mm. between its end and the magneto. Set the timing lever to position of full advance, and rotate the magneto by pedalling the machine. If a powerful spark passes regularly between the magneto and the end of the wire, it is clear that the magneto itself is in working order, and the fault must be looked for in the cable or sparking plug.

If the fault is in the cable, it may be detected by attaching the latter to the magneto in the manner provided, and holding the opposite end so that the brass terminal piece intended for connection with the sparking plug nearly touches the cylinder or other metal object. Now again rotate the engine by the pedals, and if the cable is all right, a spark will jump across the gap between the terminal and the cylinder. If it does not, there is something which requires remedying, and most likely it will be a loose terminal, broken or frayed wires, or the insulation of the cable may somewhere in its length have been destroyed. Next, to test the sparking plug, attach to it the cable and adopt the same method as when testing the cable. If a good spark jumps from plug to cylinder, well and good ; if not, examine plug and find out what is wrong. Perhaps one of the points is touching the central electrode, thus short-circuiting the current ; or a speck of foreign matter has become lodged between the two, thus bringing about the same result ; or, the porcelain insulation may be broken and loose, the points, covered with oil, or other derangements have arisen. It is generally possible to repair the mischief, but the best plan by far is to change the plug for another (no experienced motor cyclist ever goes out without a spare plug at his disposal) and defer the setting of matters to rights in the affected plug until a more convenient occasion. Sometimes—in very wet and dirty

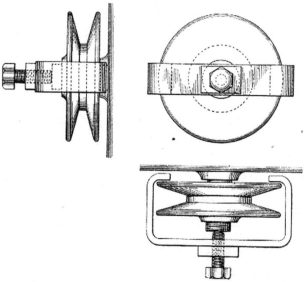

FIG. 175.—An easily-made and efficient Pulley Shifter.

weather—the working of the magneto may be interrupted by water gaining access to the interior of the machine ; but, fortunately, this does not often happen. Short-circuiting also occurs at times as a result of wet on the high-tension terminal and cable connections ; but where the magneto is carried at the back of the cylinder under the tank or a protective cover is used, these things cannot very well happen. Neither can they where the water-tight carbon-holders are in use, and those who are not fortunate enough to possess these may secure reasonable immunity by liberally smearing the terminals with vaseline in wet weather. A stuck-up A.O.I. valve will prevent the engine starting, and the plunger provided for the purpose should be depressed with the finger to free the valve from its seat.

Testing the Ignition Adjustments.

The reader's attention has already been called to the adjustments necessary to ensure a suitable timing of the magneto spark in relation to the movements of the piston, and the drawing Fig. 75, on p. 111 of the chapter on Ignition, was designed to assist him in undertaking the task for himself. When the timing of the ignition has been completed it serves a useful purpose (as a test) to run the engine on the stand and note the exact results produced by working the timing lever on the tank very gradually from position of maximum advance to that of maximum retard and the adjustment is correct when the engine will continue regularly, but slowly, firing at full retard, and the throttle is set in normal position for running full advanced on the level under load—*i.e.*, about one-third open, with air to suit. It will then be found, with most engines, that full advance can be given without any fear of causing the engine to knock, and as the sparking lever is brought slowly from retard to advance, the speed of the engine will rapidly increase up to the last fraction of movement, and it is not necessary to open the throttle wider to get the engine to continue firing when maximum retard is reached. Should the engine run faster when *almost* fully advanced than when *actually* fully advanced, it shows conclusively that too much advance has been given and the gear or chain drive must be set back slightly in the opposite direction to that in which the contact-breaker of the magneto revolves—that is, in the direction of retard.

The Engine Pulley.

The pulley is secured on the engine shaft by means of a key and keyway, and also by nuts on the end of the shaft. If these nuts work loose the full brunt of the work of retaining the pulley in place falls on the key, and should the nuts fall off on to the road the pulley will not be long in following them. These nuts should be occasionally tested with a spanner to ascertain whether any tightening up is required. The pulley may be removed from the engine shaft after removing the nuts by placing screwdrivers or wedges between it and the crank-case and then dealing some smart, but not heavy, blows with a hammer on the end of the shaft. Leave a nut slightly overhanging the shaft when doing this or else hold a block of hard wood between hammer and end of shaft to avoid injuring the screw threads on the latter.

The Brakes.

The brakes on a motor cycle may be of several different types. The law insists on each machine having two brakes, each capable of locking the wheel, and the majority of up-to-date machines are equipped with a brake on each wheel, that on the front one being applied by Bowden mechanism and acting on both sides of the rim, while the rear wheel brake is foot-applied and acts on the belt rim. Some makers employ a band brake on one side of the rear wheel and a

FIG. 176.—Band Brake for the Rear Wheel of a Motor Bicycle.

rim brake on the other; but the better plan is to distribute the braking effect over both wheels. Particular attention should be given by every motor cyclist to the condition of his brakes, as failure on the part of either may bring about an accident, while, should both be in a defective condition, the consequences may prove fatal either to the rider or some other user of the highway.

Care of the Bicycle Parts, etc.

The wheels, pedalling gear, free-wheel, steering head, and other " bicycle " parts should receive regular attention, as well as the engine, their lubrication being regularly attended to, and any adjustments necessary should be made without undue delay. The tyres require looking after, bad cuts should be stopped with a tyre filling, and an occasional look

round the outer covers may lead to the discovery and removal of what might otherwise provide a means of punc-
turing the tyre during the next ride. Loose spokes should
be put right as soon as possible after detection ; and, in
short, a careful watch should be kept on every part of the
construction. Keep the *outside* of the engine as clean as
possible, using a stiff brush dipped in paraffin or waste petrol,
so that accumulations of mud and grease can be removed
from the most inaccessible parts. A dirty, neglected-looking
engine is always a bad recommendation to the user, and,
what is more, the cooling is in a measure retarded. When

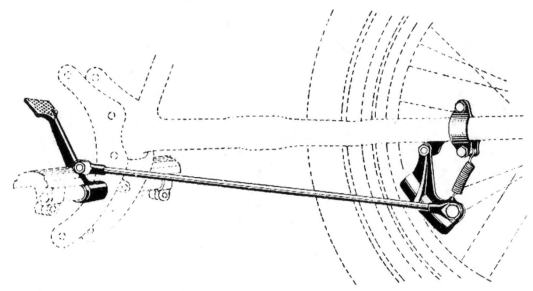

Fig. 177.—Belt Rim Brake for Motor Cycles.   A popular and efficient
type.

a grease-covered crank-case gets thoroughly warmed up
the odour given off from it is a very unpleasant one.

Methods of Starting.

With each and all of the foregoing engine adjustments
properly attended to, there should be no difficulty in starting
at the first attempt.   It is advisable, especially during cold
weather, to inject a little paraffin and petrol mixed through
the compression tap into the cylinder, so that the piston rings,
between which and the cylinder walls a film of thick oil is
deposited when the engine is stationary, may be freed from
their " gummed up " position.

To effect starting of the engine, apart from the machine

as a whole, the back wheel should be jacked up off the ground by means of the stand provided for that purpose, the petrol turned on, carburettor flooded by depressing the plunger on float chamber, spark lever well, but not fully, advanced, and the throttle opened from one-third to one-half with the extra air either closed entirely, or, in some cases, slightly opened. Raise the exhaust valve lifter and smartly pedal the engine round, dropping the exhaust lifter directly a fair speed has been got on the engine—*i.e.*, after two or three turns of the pedals. The engine should at once fire and, if the owner is a novice, he will be well advised to acquaint himself with the manipulation of the various levers, etc., before venturing out for his first ride. On no account whatever should he keep the engine running for more than half a minute at a time on the stand, as, in these circumstances, it is deprived entirely of the cooling effects which it obtains when out on the road, and will get very hot if made to continue running under such conditions. A pumpful of oil may be given with advantage during the stationary test.

CHAPTER XII.

THE MACHINE AS A WHOLE—THE SELECTION AND PURCHASE OF NEW AND SECOND-HAND MACHINES—DRIVING HINTS —PASSENGER MOTOR CYCLES—THE COST OF MOTOR CYCLING.

IN the preceding chapters the author has endeavoured to explain the construction and functions of the various parts of a motor cycle and the engine which propels it. Also the principal adjustments necessary to keep the machine in proper working order, and so forth, and by this means it is hoped that the reader will have acquired sufficient insight

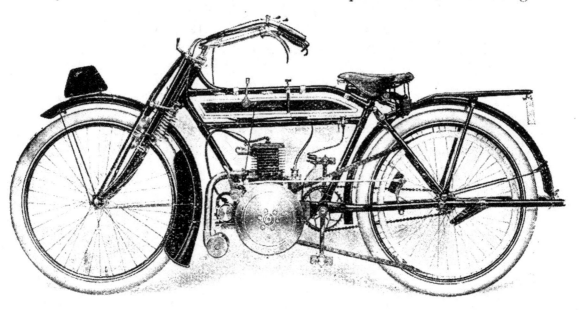

FIG. 178.—The "Connaught" Two-stroke Motor Bicycle, 2½-3 h.-p.

into the subject to enable him to discern what is requisite in the completed machine, and to be able to judge somewhat of the differences which exist between one type and another.

The average purchaser of a motor bicycle has a fairly big proposition to deal with, unless, for one reason or another, he be possessed of means or is able to decide off-hand exactly which make of machine he will favour. The man who, knowing little or nothing of the subject, decides to become a motor cyclist, is, as a rule, led first one way and then the other by the advice of friends or the blandishments of the makers' representatives ; and in the end it may quite well come about that he buys a machine, not so much because of his own inclinations in the matter as because of the wonderful qualities others have assured him it possesses. He may quite possibly profit by the advice he gets, and secure thereby a motor cycle which will serve him faithfully and well for a very considerable time ; but, on the other hand, he is just as likely to acquire either an unsuitable or an inferior make, and be led by the consequences to give up the pastime altogether.

The prospective motor cyclist should, therefore, look well into the matter before he parts with his money — by all

FIG. 179.—The Lady's 2½ h.-p. " Hobart."

means consulting those who know more about it than he does—but reserving the summing up and final judgment to himself after mature deliberation. A motor bicycle costs a fair sum of money, and to suitably equip it with accessories, pay licensing and other fees, and generally set up as " a motorist in a small way," and then discover that another make would have met his particular requirements much better, is, to say the least, disheartening, and likely, as before intimated, to make one " give up the game " altogether.

The Question of Type and First Cost.

The first thing to be considered is—what can the intending purchaser afford to expend, and, that decided, which *class* (as apart from make) of machine will best conform to

his requirements?   If the beginner is of moderate physique
and pecuniary resources, he may do best to invest in a good
second-hand light (or medium) weight machine, with engine
power ranging from 2 to 3 h.-p., which will take him almost
anywhere he is likely to want to go, and do the work cheaply
in respect of petrol and oil consumption, wear of tyres and
general upkeep.   If, however, he be fairly well set up in both
respects and entertains a preference for something more
powerful, a $3\frac{1}{2}$ h.-p. bicycle with heavier tyres, greater weight,
and costing rather more to maintain, will, perhaps, be more
suitable for all-round use.

Where expense is not such an important matter and the

FIG. 180.—" Humber " Twin-cylinder $2\frac{3}{4}$ h.-p. Motor Cycle with Three-
speed Gear.

rider can afford to pick and choose from among the best
motor cycles, he will doubtless make his choice of a brand
new machine, with such refinements as may appear to him
desirable.   It depends very much upon the nature of the
work to be performed, the financial position of the purchaser,
his temperament and capabilities.

If a lot of heavy work is to be done—*i.e.*, riding in difficult
country with a bulky outfit, going out in all weathers and
wherever business or pleasure takes one—it is advisable to
have power and plenty of it ; and to obtain really good
results it may be desirable to purchase a machine fitted with
a free-engine clutch or two-speed gear.   It is difficult enough

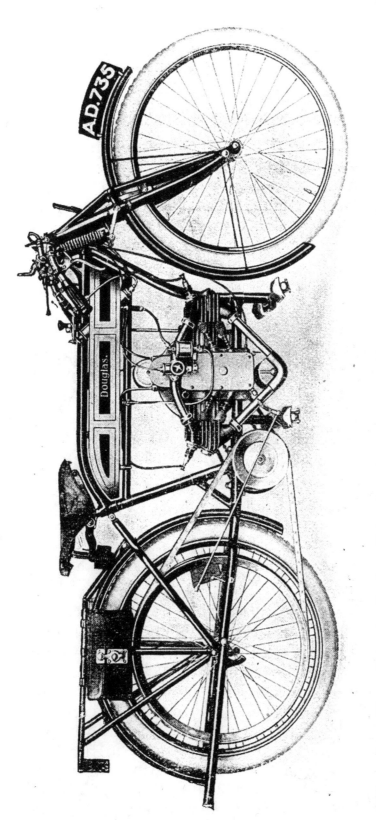

FIG. 181.—The "Douglas" Motor Bicycle, with $2\frac{3}{4}$ h.-p. Engine having Horizontally Opposed Cylinders, Two Speeds and Free Engine.

to offer advice in such a matter as this, for there are so many individual preferences and considerations to be taken into account that it becomes almost impossible to lay down any system, let alone a definite law, to be followed by the novice, or, for the matter of that, by the more experienced motor cyclist either.

Avoid Unnecessary Weight and Excessive Engine Power.

One thing, however, may be urged with conviction, and that is, that no one, not possessed of average activity and strength, should burden himself with a heavy, high-powered, complicated machine, far in excess both as regards engine capacity and weight of his requirements, and needing an extra lot of looking after to keep in good order. At

FIG. 182.—The "Regal-Green" Motor Bicycle, 3½ h.-p., Water-cooled Engine. (See also Figs. 11 and 11A, pp. 20-22.)

all other times than when the rider is being carried along by it, and sometimes even then, the handling of such a machine inflicts a heavy tax on the resources of its possessor. The strong and youthful, or comparatively youthful, reader will probably disagree with these remarks and contend that there is "nothing like" a 7 h.-p. twin, which will roar up any hill you like to put it at and jib at nothing to be met with on the roads of this country ; but in time he, too, will come round to a saner way of viewing the matter and be content with a lighter, cheaper, and altogether handier mount, which can be used for almost any purpose and worked within a reasonable distance of its power capacity most of the time, instead of some 50 per cent. of the engine power developed going to waste practically all the time, besides other disadvantages.

Where the intention is to employ a side car and take an extra passenger *frequently*, and the districts to be traversed include some stiff hills, it is not only advisable, but indispensable, to employ a powerful engine, set in a strong and heavy frame, and, further than this, to utilise a variable speed gear, and if (as before remarked) real touring and the adoption of the " go anywhere and do anything " principle is contemplated ; but, unless this be the case, the purchaser will be much better advised to modify his tastes and go in for something of lighter build.

The beginner may well be pardoned if he fails to understand how it is that so many engines, with differently proportioned cylinders, can be described as of the same horse-power. His best plan will be not to worry about the point, for practically all motor cycle engines develop a greater horse-power than the makers claim for them.

The R.A.C. (Royal Automobile Club) has provided a simple formula for calculating the approximate horse-power of motor-car and cycle engines, as follows :—

$$\text{R.A.C.} \qquad \text{H.P.} = \frac{D^2 \times N}{2 \cdot 5}$$

when stroke is taken into consideration.

$$\text{H.P.} = \frac{D \times S \times N}{2 \cdot 5}$$

where D = diameter of cylinder in inches.

,,   S = stroke of piston in inches.

,,   N = number of cylinders.

And to ascertain the cubic capacity of the cylinder the most simple plan is to square the bore of cylinder × ·7854 × stroke.

Second-Hand Motor Cycles.

It is a good plan for the absolute novice to commence operations by purchasing a second- or third-hand machine, with which to serve his novitiate as a motor cyclist. A really good and fairly modern machine can be picked up for a moderate sum—say, £16 to £20, and many for considerably less than this, being of older pattern ; but it is necessary that the beginner should go exceedingly carefully to work, or he may find himself landed with a machine of little worth until a further sum has been expended in putting it into working order. If he numbers motor cyclists among his friends, he should prevail upon one or more of them to go along with him to inspect the proffered mount ; but, if this is impossible, he may adopt one or two simple precautions as a guard against being misled by an unscrupulous, or too enthusiastic, vendor.

Inspection before Purchase.

The compression of the engine should be tested by standing on the pedal, with rear wheel jacked up clear of the ground, and at least several seconds should elapse before the weight of the person making the test forces the engine round

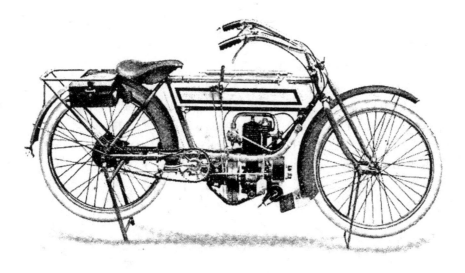

FIG. 183.—The "F.N." 2¾ h.-p., with Shaft and Bevel Drive, Two-speed Gear and Clutch.

over compression. If there are no pedals, wheel the machine along the ground with belt on and exhaust-lifter released, when, if the compression be in order, the back wheel will skid as the piston travels upward towards the end of the compression stroke. Examine the pulley shaft for wear, taking hold of the pulley itself and seeing whether there is any *vertical* movement. If there is to any appreciable degree, it may be taken as evidence of the fact that the engine shaft, and most probably other bearings, also require re-bushing, and this should weigh very considerably in the mind of the prospective purchaser and go far towards deciding him whether or not to go any further in the matter.

The condition of the ignition apparatus, whether accumulator or magneto system, should be closely investigated, and anyone who has taken the trouble to read the foregoing

FIG. 184.—The "Rudge-Multi," with Infinitely Variable Gear.

explanations in the Chapter on Ignition, should be able to tell within a little whether these appliances on a second-hand motor bicycle are in reasonably good order or not. Next, it

is advisable to inspect the general state of the engine, noting
whether any part of the radiating fins have been broken off
the cylinder, or other signs of rough or negligent treatment
exist. If the previous owner has been careless or unfortu-
nate in these respects, the still more important interior parts
may have suffered equally in his hands, and the exterior
condition of the machine may generally be taken in some
respects as an index of the condition of affairs inside.

The frame, wheels, and other " bicycle " parts should be
looked over, loose spokes or other likely faults, if possible,
detected, and pointed out to the seller, while the make and
condition of the tyres should be especially noted. The price
asked should depend in some measure upon what accessories

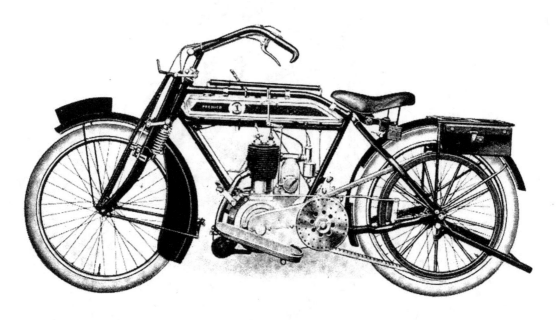

Fig. 185.—The " Premier " Two-speed Motor Cycle, of $3\frac{1}{2}$ h.-p.

are included in the bargain, how old the machine is, and
what amount of mileage it would appear to have covered.
The owner's estimate of this latter point, plus (in most cases)
some 50 per cent., may be taken as representing something
near the truth.

Above all, insistence should be laid upon facilities being
afforded either for a personal test or an exhibition of the
powers of the engine on the level and up a steep grade.
Very little heed should be taken of the owner's statements
if the machine fails on this occasion, under fair and reasonable
able conditions, to do what is required of it.

The Vendors of Second-hand Motor Cycles.

Perusal of the foregoing remarks may, perhaps, lead the reader to suppose that vendors of second-hand motor bicycles as a class are prone to be dishonest, and that the inexperienced buyer is running a great risk in dealing with them. This, however, is not at all the impression which it is desired to convey. There are several firms, both in London and elsewhere, who do a large and perfectly straightforward business in this class of goods ; and not only is it possible to obtain exceptional bargains from them at times, but one may always rely on receiving the expert assistance of a

FIG. 186.—The " Service " Touring Motor Cycle, 3½ h.-p., with Three-speed Gear and Free Engine.

thoroughly competent and experienced salesman in making a selection.

The same applies to the private advertisements one sees in such profusion in the advertisement columns of the Press. Probably nine-tenths of the machines therein offered for sale are worth what is asked for them, and often more perhaps ; but there remains the odd tenth, and it is because of this fact that readers are warned to be as careful as they possibly can in purchasing a second-hand machine.

Purchasing a New Machine.

The fortunate individual who can afford to pay cash down for a brand new mount is relieved of a lot of worry in that he has a host of good firms' productions to choose from, and he may take it that the condition of the machine will be all that can be desired when it is delivered to him. It is important that the order should be placed as early in the year as possible, for the makers, none of them, go in for turning out large numbers of any one pattern on chance, but work according to the run of orders, so that as long as from eight weeks to three months often elapses between the date of the order and that of delivery. Lest this statement be contradicted, the author might, perhaps, quote the experience of a personal friend. Four prominent makers, one after the other, required the order to be placed early in January, for delivery by Easter in two cases and Whitsun in the other two, of that year. This is, perhaps, an extreme

Fig. 187.—The " P. & M." 3½ h.-p. Motor Bicycle, with Chain Drive and Two-speed Gear.

case, but it is generally several weeks before one can obtain one's new machine.

Having once acquired a motor bicycle, and learned how to drive and manage it, a new interest is created in the life of the possessor. No longer need he depend on the railway time-table when desiring to make a journey of any length. It is often not only as quick, but actually quicker, to go by road, and how much greater is the enjoyment of so doing— given, of course, reasonable weather and other conditions. The fact is brought home with considerable force where the journey is a cross-country one, and one, or perhaps two, changes have to be made *en route*, and the author has himself frequently scored over the railway under these and similar circumstances.

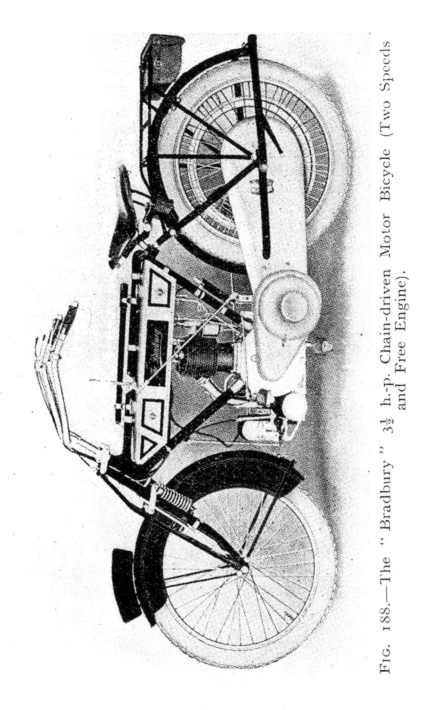

FIG. 188.—The "Bradbury" 3½ h.-p. Chain-driven Motor Bicycle (Two Speeds and Free Engine).

Various Types and their Capabilities.

The motor cycle is essentially a pleasure vehicle, but at the same time its capabilities are such as to render it an invaluable aid to commercial travellers, doctors, road surveyors, and others whose work takes them out and about at all times of the day ; and a point worth mentioning here is that where frequent stops and re-starts have to be made it is advisable to use a lightweight machine (for town and runabout purposes), and a clutch or two-speed model or cycle car where longer distances and heavier going have to be provided against.

Always remember that some day or other you will be wanting to sell your motor bicycle, so go in for as good a one as possible, and take care of it *all* the time, so that a better price can be obtained when the time comes for selling.

It would occupy far too much space to set forth even briefly the leading features of the motor cycles at present on the market, and it should be clearly understood that, in selecting for illustration and special notice those of which reproductions are hereby appended, the purpose has been to convey a true impression of the different *types* of machines, irrespective of any particular claims made on their behalf by the manufacturers. The whole book might be easily filled with illustrations and particulars of other models equally as deserving of notice.

Desirable Features of Construction.

The purchaser will doubtless be guided to a large extent in making his selection by his own personal preferences, and will eventually decide on the machine which embodies in its general design the largest number of points which accord with his own views ; but it may, nevertheless, be of use to the less experienced buyer to have placed before him a few of the essential points which he should bear in mind when endeavouring to come to a conclusion as to which machine he will have. Briefly summarised, these may be taken as follows :—

The machine should possess :—

(1) An engine, in the design of which accessibility to every part has been studied and provided for.

(2) A soundly designed and strongly made framework, with the engine so carried as to be easily detachable, if desired.

(3) A low-riding position, with a comfortable reach and adequately designed foot-rests or, preferably, foot-*boards*.

FIG. 189.—The "Scott" 3½-4 h.-p. Motor Bicycle, with Two-stroke Twin-cylinder, Water-cooled Engine and Chain Drive.

FIG. 190.—The "Stellar" Two-stroke Twin-cylinder Motor Bicycle, with "Stuart" Water-cooled Engine and Worm Drive.

(4) Spring forks, which absorb vibration without permitting " bouncing " of the front wheel or lateral instability.

(5) Handle-bars having an easy rake, adding to the comfort of the rider and his control over the machine.

(6) A fairly capacious tank.   (This, of course, depends to a large extent upon the engine power and weight of machine.)   The tank-fillers should be large in diameter, especially that of the lubricating compartment.

(7) An efficient silencer (*not* an apology for one).

FIG. 191.—The " Clyno " Motor Bicycle, 5–6 h.-p., Three-speed Gear and Chain Drive.   An ideal side-car machine for heavy work.

(8) First-grade tyres, suitably proportioned for the work to be placed upon them.

(9) A belt of adequate width.   If chain-drive, only the best makes of chain are any use.

(10) A complete tool outfit, stand, carrier, and number-plates (these are *not* considered accessories now-a-days by good makers).

(11) A decent saddle ; and

(12) Good finish all round.   Magneto (which should be placed in a protected position, as at the rear of the engine) and carburettor of known and well-tried reputation.   General handiness, first-class

enamelling and plating, lack of fancy work, and no unnecessary complication of detail.

The purchaser should not hesitate to enquire as closely as he likes into the hidden characteristics of the engine.

FIG. 192.—The " Rex " Open Frame Motor Bicycle. Belt Drive, Two-speed gear, 6 h.-p. Twin-cylinder Engine, Handle Starting.

Many leading firms publish in their catalogues interior views of the engines and other parts of the construction ; but others leave the purchaser to find out from experience the features of design and class of workmanship and material he has been supplied with ; and the first time he has the engine to pieces he may find evidences of faulty design, hurried workmanship, cheap labour, or other drawbacks which, had he known at first, would have decided him against purchasing the machine.

The Beginner's First Ride.

It is wonderful how soon a complete novice falls into the ways of motor cycling—almost at his first ride, as a rule. He should take the machine out on a quiet road and, if possible, find an easy down grade whereon he can pedal the machine without the belt, in order to get used to the position and control. This done, he can next have a try with the engine running, making sure to keep the exhaust

FIG. 193.—The " Ariel " 7 h.-p. Motor Bicycle.

lifter and brakes handy so as to stop immediately, if occasion arises.

The " running mount " indulged in by the majority of riders should not be attempted at first, although it is easily accomplished after a little practice. The exhaust valve must be raised and the machine pushed along the road at a fair speed. Then drop exhaust lifter, and directly the engine fires jump into the saddle, either by means of the pedal or straight off the road, immediately afterwards setting the levers as described below. As he becomes more experienced the rider will know just how to set his levers to get going easily, and his knowledge of the machine and its ways will rapidly extend. Something will be learned every time he goes out, and it will not be long before he feels himself competent to go almost anywhere and tackle almost anything that the more seasoned and experienced motor cyclist undertakes.

Driving Considerations.

The great thing in driving a motor cycle is, of course, to humour the engine as much as possible. Keep the petrol consumption down by running with the throttle as far closed as circumstances permit, and the air lever, on the contrary, as wide open as possible. Petrol costs money, air does not ; therefore, study your own pocket by husbanding the former at the expense of the latter.

When approaching a stiff hill get all the speed on the

Fig. 194.—An American Motor Bicycle : The " Indian " 7 h.p. Chain Drive and Two-speed Gear.

engine possible, and begin by having both the throttle and air levers well open. Then, as she shows signs of flagging a little, give the remaining throttle, and place your finger on the air lever, slowly closing it as becomes necessary to maintain the power of the engine, by feeding it with a richer mixture. Further signs of flagging should be met by closing the extra air still more (and perhaps entirely), and then, when that is done and something more is wanted, try slightly retarding the spark lever so as to cause the ignition to take place rather later, viz., when the piston is in such a position and the crank at such an angle as to be more favourably

disposed to receive the impulse.   If pedalling gear is fitted use should be made of it if the engine begins to labour, and before, *not* after, signals of distress have become apparent.

Do not be afraid to try the effect of *reopening* the air lever (after shutting it off as described above), by slight degrees, as the machine nears the top of the grade ; or, in any case, when it appears to be mastering its task.   When it has had a good feed of rich mixture it often calls for a slightly weaker one, something it can digest better, and a wonderful spurt follows the reopening of " the air."   Lose no time, on

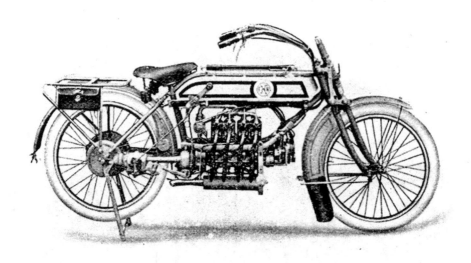

Fig. 195.—The " F.N." 5-6 h.-p. Four-cylinder Motor Bicycle, with Bevel Gear Transmission.

reaching the summit, in closing the throttle down to its normal position, and if the climb has severely tried the engine, it is as well to run at moderate speed for a bit.   The spark lever may now be advanced, but never while the climb is in progress.   Every opportunity which presents itself for cooling the engine should be taken advantage of.   When running down a grade, lift the exhaust valve, open the air lever wide, and partly open the throttle.   Coast as far as possible, allowing the cold gas to circulate through the engine, and if a switch is fitted, use it for the purpose of suspending the sparking operation until the engine is again required. Do not let the speed get down too low before restarting the engine.

" Misfiring " occurs when too much extra air is given,

for the reason that the mixture is then too weak to be exploded with the necessary force ;   also, if there is any appreciable impediment in the petrol supply or the insulation of the sparking plug is defective.   If any part of the high-tension cable wire should become exposed and touch any metal object, the current will be interrupted and misfiring set up.   A sticking valve or broken valve spring may cause it, and so does the presence of dirty oil on the sparking plug or contact-breaker points.   The rider cannot remain long in ignorance of the fact when his engine is even slightly misfiring.   The speed falls off and the noise of the explosions becomes fitful and altogether irregular.   The sparking plug

FIG. 196.—The " T.M.C." Four-cylinder Motor Bicycle with Bevel Drive and Three-speed Gear.

is usually found to be at the seat of the mischief when misfiring occurs.

Care as to Lubrication.

The continued use of a rich mixture (too much gas, too little air) is not only uneconomical, but tends to overheat the engine, destroying lubrication and generally acting adversely towards it.   A pumpful of oil injected into the engine as the hill is approached helps wonderfully, and in every condition of running the lubrication of the engine must —as was emphasised in Chapter IX—be strictly attended to.

The amount of oil required varies with the size and speed of the engine. A standard $3\frac{1}{2}$ h.-p. engine, doing solo work with an average weight of rider and luggage, requires a full pumpful of oil about every 10 to 15 miles. More when it is quite new, when it is working under adverse conditions of road and weather, or when prolonged running in a low gear is enforced, as when passing through thick and continuous traffic, and the engine is kept going at a high speed by means of a clutch or variable speed arrangement, the cooling of the

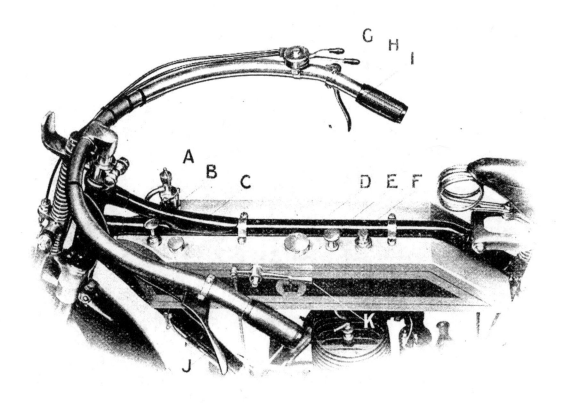

FIG. 197.—Control, etc., Fittings of the " Rex " Motor Bicycle.

(A)—Automatic Lubricator. (B)—Oil Pump. (C)—Oil Tank Filler. (D)—Petrol Tank Filler. (E)—Petrol Injector. (F)—Petrol Gauge. (G)—Carburettor on Lever. (H)—Throttle Lever. (I)—Front Brake. (J)—Exhaust Lifter Lever. (K)—Ignition Lever.

engine being curtailed at the same time by the slow movement of the machine through the atmosphere.

A better method than giving a full charge of oil every so many miles is to give half-a-pumpful twice as often, as then the lubrication is effected with greater continuity—not a surplus at one time and a shortage at another. Much depends on the size of the oil pump and the quality of the oil employed.

Engine Control Methods.

As regards which is the best method of controlling the engine on the road, the reader is recommended to cultivate the throttle and control method. This tends to economy in petrol consumption, certainly in results, and without doubt produces a greater efficiency all round. Raising the exhaust-lifter, when desiring to reduce speed, is all very well, but it is hard on the exhaust valves, uneconomical, and is generally called "bad driving." Switching on and off of the electrical current provides another means of control; but it is not a good one, and the engine is subjected to shocks which it might be spared by employing the throttle control system.

Either of the last-named methods conduces to the production of what is known as "popping" in the silencer, a thing which does much towards bringing motor cycling greatly into disrepute among the general public. Gases are allowed to continue circulating through the engine, and they reach the heated silencer in an unexploded state. The high temperature in the latter does the rest. The throttle method cuts down the gas, and entirely bars its passage to the cylinder when not required, and there is practically no risk of explosions taking place in the silencer in such circumstances.

The Question of Gear Ratio.

The gear ratio of a single-geared motor cycle plays an all-important part in the working of the engine. For speed work a high gear is required, and a lower one for heavy pulling and hill-climbing work. The means whereby these opposite interests are combined in the one machine have already been described, but it remains to be said that many "fixed" gear motor cycle engines are undoubtedly over-geared. The reader may require a little guidance as to what constitutes a suitable gear ratio and how he may ascertain what his particular machine is geared at.

To ascertain this latter point a small file or other mark should be made on the engine pulley, and another similar mark placed on the belt rim of the back-wheel. Get both in the topmost position—over the centre they revolve round. Then (preferably with the help of a friend) turn the back wheel one complete revolution, and note how many revolutions the pulley has made during that time. By this simple means the gear of any belt-driven machine may be at once

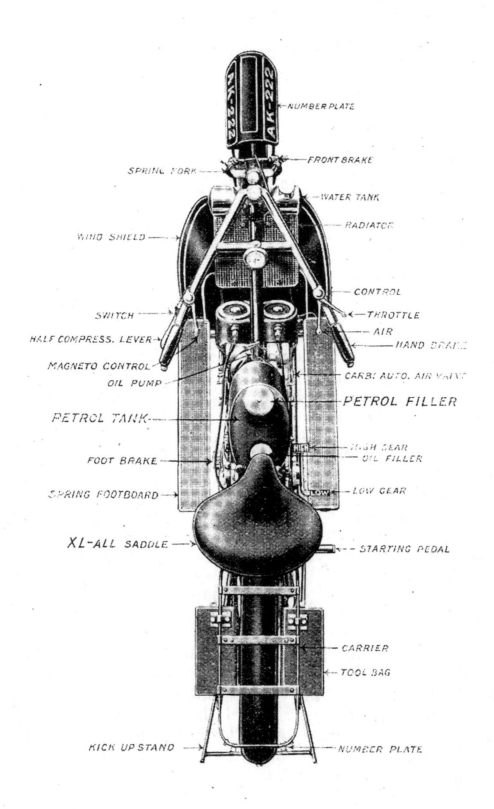

NUMBER PLATE

FRONT BRAKE

SPRING FORK

WATER TANK

RADIATOR

WIND SHIELD

CONTROL

SWITCH

THROTTLE

AIR

HALF COMPRESS. LEVER

HAND BRAKE

MAGNETO CONTROL

OIL PUMP

CARB: AUTO. AIR VALVE

PETROL FILLER

PETROL TANK

HIGH GEAR

OIL FILLER

FOOT BRAKE

SPRING FOOTBOARD

LOW GEAR

XL-ALL SADDLE

STARTING PEDAL

CARRIER

TOOL BAG

KICK UP STAND

NUMBER PLATE

FIG. 198.—The "Scott" Motor Bicycle from above.

ascertained ; the number of pulley revolutions compared with the single revolution made by the belt rim, of course, constituting the gear ratio of the machine.

For ordinary touring work, taking the hills as they come in average country, a gear of $4\frac{1}{4}$ or $4\frac{1}{2}$ to 1 will suit an 11-stone or heavier rider, with luggage in addition ; but where heavy country has to be traversed, and the load is much increased, better results will be obtained with a gear of, say, 5 to 1, while for temporary purposes even 6 to 1 may be used with advantage.

An adjustable pulley will usually provide a range of gears such as this ; while if a variable speed gear is employed, it is possible, of course, to obtain a 9 to 1 gear, if required.

Riders of moderate weight in plenty are employing gears of less than 4 to 1 and getting on very well with them ; but one has to be very active and a good driver unless a lot of engine knocking is to take place.

" Knocking "—What it Means.

The term " knocking," referred to in Chapter VII, describes the clanking, metallic noise sometimes heard within the cylinder, and caused either by reason of the gear being too high, heavy carbon deposits in the cylinder, worn bearings, or faulty driving, with spark too far advanced at low speeds. To correct it the extra air-opening must be reduced at once, and if the knocking is very persistent a lower gear must be employed, the ignition set back a trifle, the piston, etc., cleared of carbonised accumulations, or the bearings rebushed, according to which of these causes is producing the mischief. In any case it means that oscillation is being set up in the bearings and hammering action produced—to the detriment of the engine generally.

The task of re-bushing an engine is, generally speaking, beyond the powers of most motor cyclists, principally owing to lack of the necessary tools. Care should be taken to entrust this work to competent motor mechanics only, and the charge should largely depend upon whether the owner or the repairer takes the engine down and re-assembles it.

Some " Don'ts " to be observed.

Summarising the foregoing remarks in a series of " Don'ts " which, even though they *do* repeat what has already been said, can hardly fail to serve a useful purpose, we would say to the reader who desires to study his own interest and that of the engine he drives—

Don't run the engine for a moment longer than is absolutely necessary on the stand before starting.

Don't attempt to start the machine under almost impossible conditions, and then grumble at the difficulty.

Don't grab at the throttle and sparking handles and alter their positions suddenly, with a jerk. The manipulation should be gradual and by small degrees, unless in the case of an emergency arising.

Don't retard the magneto on hills unless absolutely obliged (to prevent knocking), and then only slowly and by small degrees.

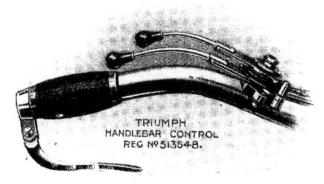

FIG. 199.—Carburettor, Handle-bar Control, "Triumph" Motor Bicycle.

Don't forget that air is cheap and petrol is not, so throttle down and run on as weak a mixture as circumstances permit.

Don't fail to take every opportunity which presents itself for cooling the engine on down grades.

Don't switch in the magneto again while running switched off, unless there is a fair speed on the engine.

Don't control the speed by frequently raising the exhaust lifter. Learn to "drive on the throttle," and rely on that method for regulating the speed.

Don't use the exhaust cut-out (if you have one) unless well away on the open road. Never open it near restive horses or when passing through the streets of villages and towns.

Don't hustle the engine unnecessarily at any time. Forced speed does harm, and one never knows when the policeman may appear from behind the hedge.

Reasonable care only, such as anyone can give, is required, and it is generally amply repaid, as not only is the efficiency of the engine and machine as a whole upheld by careful treatment, but there is a certain indefinable pleasure in looking well after one's mount, and the fact that, after all, it is a somewhat complex piece of engineering construction should never be quite lost sight of.

What the Tool Bag should Contain.

When setting out for anything in the nature of a *long* ride the motor cyclist should take with him the following kit of tools and " spares " :—

A large adjustable spanner ;

A smaller-sized ditto ;

The set of spanners and any other special tools supplied with the engine ;

Screwdriver ;

Pair of pliers (combination type) ; another pair (quick-grip type) ;

Belt-punch and knife (or small saw) ;

Some copper wire, high-tension cable, a few assorted bolts, split pins, nuts and washers ;

Two or three spare sparking plugs, a few spare magneto parts (such as high-tension carbon holder), platinum points and contact-breaker bell-crank, insulating tape and a file for touching up platinum points ;

Spare belt, a couple of belt fasteners, butt-ended inner tube for tyre, repair outfit for ditto, tyre levers, and some valve rubber ;

Complete inlet and exhaust valve for engine ;

Spare carburettor jet ;

Carbide (and perhaps a spare lamp burner.)

Of course, if only an ordinary pleasure outing, covering a comparatively short distance, is intended, many of the above items may well be left behind ; but the motor cyclist of average carefulness usually carries at least the outfit listed above when going any distance, and many take nearly twice as much. One may be laughed at for carrying a " blacksmith's shop in miniature " about with him. The author was on one occasion subjected to some good-humoured bantering of this description, and shortly afterwards met the banterer-in-chief in a remote part of the country late at night and stranded for want of a simple part, which was at once produced and thus saved what would have otherwise have meant, for the erstwhile joker, a four-mile trudge pushing a heavy machine.

These remarks must not, however, be taken as implying that it is necessary when motor cycling to cart a lot of cumbersome tackle about with one. Moderation is required in

FIG. 200.—The "New Hudson" Three-speed Motor Cycle and Side Car.

this as in all things ; but an article or so too much is better, when the pinch comes, than having to suffer delay and inconvenience because of some " missing link "—perhaps not worth more than a few pence when the next town is reached —but cheap at any price when out on the lonely road.

Motor Cyclist's Clothing.

As regards suitable clothing for motor cycling. In this country, at all events, it is hardly ever safe to venture out without some sort of mackintosh outer covering. Stout boots, warm underclothing, and a suit and cap of thick material are required for ordinary weather riding ; while in winter, or for night journeys, a leather waistcoat, leggings, and a good warm neck-wrap add to one's comfort. The mackintosh overall suit should be of double texture, and the jacket double-breasted. Leather gauntlets or some other heavy class of leather gloves are essential. In wintry weather the author uses a Service "Coldweather" jacket made of lined Irish freize, with wind cuffs and storm collar ; leather breeches, stout leggings, and boots, with, perhaps, a leather waistcoat and sweater into the bargain ; and finds this combination very efficient in keeping out the most intense cold. In the summer months holland overalls are useful, but, as before said, it is never wise to go out without mackintosh protection as well. The Burberry motor clothing has special advantages of its own.

FIG. 201.—Side Car with Enclosed Body and Side Door.

Some Hints " In Lighter Vein."

It is unwise to keep the machine at it for just so long as it will run, without ever troubling to look into the condition of the parts from time to time. Be sure that if this negligence is indulged in, the time will come, as certain as Fate, when she will let you down out on the road, and it will be miles from anywhere, for these things never happen just outside the wide-open doors of a motor garage.

Never imagine you know *all* there is to know about motor bicycles. The designers themselves don't—in spite of their wide experience, and the author of these pages realises only too well what a lot *he* has yet to learn about the subject. Let every ride you undertake teach you something, however small. Join a club or motor cycle organisation of some sort, and go where motor cyclists "get together," and keep your ears open. You will, maybe, hear a certain amount of nonsense talked, but there will be plenty of sensible and instructive conversation also, and much of it you can turn to your own advantage in a practical way. Always know you may

FIG. 202.—The "Scott" Motor Bicycle and Coach-built Side Car.

be stranded one day (or night) where assistance is entirely unavailable, and something has gone wrong with the machine which you cannot fathom at first. You will then have to extricate yourself from the situation off your own bat, either by triumphing over the difficulty, or by hard pushing till some more favoured spot is reached.

The Man or the Machine?

How many times does it occur that motor cyclists are held up on the road for hours together, when only just a *little* more knowledge of the construction of their machines and the adjustments needed to keep them in order would have enabled them to get going again in a fraction of the time. It must always become a triumph of mind over matter in these circumstances. The machine must be made to go where the rider wants to go, and do what he wants it to do, and *not* the other way about, and when she stops without his permission there is something which has to be discovered and put right. Motor cycles don't stop for nothing or just out of pure fun ; there is always a reason for it, and as they can't tell you what that reason is themselves, you have got to discover it and take the necessary steps for its removal. It is part of the game, and a very fascinating part into the bargain. Don't get easily disheartened, but put on your considering-cap, have a smoke, and go methodically to work.

There will be generally some more or less pronounced symptom to guide you, but if not there is no cause for alarm. Don't go spending an hour trying to hunt out the cause of failure unless you either know, or have first made sure, that there *is* petrol in the tank. The engine simply won't run without spirit, but many a motor cyclist has discovered, after an hour or so of fruitless endeavour, that he has been trying to make it do so.

" Shall I——" ?

Should another motor cyclist overtake you while you yourself are travelling at the speed which best suits your purpose, don't jump to the conclusion that he is necessarily challenging you to a race ; but let him go on and be the one to run into the arms of the waiting policeman. Never look condescendingly at the pedal cyclist as you pass him struggling up the hill, while you are simply roaring up. The sparking plug may give out the next moment, and then he has the laugh of you.

Always be ready to help a brother motor cyclist in a difficulty. When you come across him attending to his machine by the roadside, ease up and call out to enquire whether you can be of any assistance. A small contribution from your toolbag may set him going again, and, remember—it may be *your* turn next.

Don't heed the witticisms of the passer-by who refers to you as "another broken-down motor cyclist," whereas, perhaps, you are only making a most simple adjustment. A certain class of pedal cyclists and others delight in doing this (and to change a sparking plug even is to be "broken-down" in their view); but wait until you catch them up (it won't be long), and then they wish they could change places with you after all. It is much more dignified than shouting out about their being "behind the times," and so on.

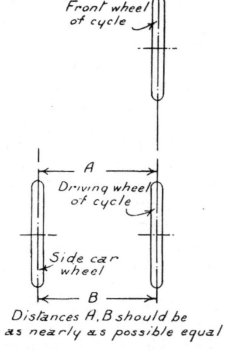

Distances A, B should be as nearly as possible equal

FIG. 203.—Method of Aligning a Motor Bicycle and Side Car.

"Passenger" Motor Cycling.

Having become sufficiently expert in the management of a motor cycle, the rider will probably yearn for some means of sharing his enjoyment with others. It seems rather selfish to be always prating to others about the delights of motor cycling, and the average man begins, as a rule, to feel that he would like to be able at times to offer a ride to his friends or relations, who otherwise may be quite debarred from anything of the kind.

Side Cars.

This may involve the purchase of another and more powerful motor cycle, but not necessarily so, and the cheapest and handiest method of carrying the passenger idea into effect is to purchase a side car. This can be attached **and** detached at will, so that either solo or passenger riding may be indulged in as fancy dictates. The author has had something like 40,000 miles of motor cycling with a side car, and finds the companionship of a passenger delightful. It is highly necessary to employ a machine of ample power and possessing a two-speed gear to secure the best all-round results ; but a great amount of " fun " can be derived from a single-geared motor cycle and side car, provided, of course, that not less than $3\frac{1}{2}$ h.-p. is available.

In his view it is by far the better plan to attach the side car to the *left*-hand side of the machine, as then in overtaking other vehicles the driver and not the passenger gets first view of the road ahead. Opponents of this view assert that as the majority of motor cyclists accustom themselves to mounting from the left-hand side of the bicycle the side car ought to be connected up on the right, so as not to preclude a continuance of the practice. They also claim that the right-hand side is better for fast riding and taking sharp corners at speed. No one worth his salt, however, need make any difficulty about mounting from the right instead of the left hand side. It is not as though he has any balancing to do.

Side Car Types.

There are two main types of side car—the rigid and castor wheel. In the first-named the wheel is held in rigid alignment ; and in the second, the side car wheel is attached to a ball-bearing steering head, and adapted to move freely and automatically in the same direction as the front wheel of the motor bicycle. The 40,000 miles above mentioned have included the use of both types, as manufactured by the well-known firm of Montgomery & Co., Coventry, who specialise in this class of work and produce a first-class article, and the opinion arrived at as a result of this is, that for all-round convenience, freedom from slipping, and simplicity, the rigid side car is the best. It is, of course, the cheapest to purchase. If properly aligned with the bicycle the wear on the tyres is but little increased, but it cannot be too urgently impressed upon prospective users of this form of machine that where it is intended to employ a powerful engine and go in for real touring or general road work the driving tyre of the bicycle should be of larger section than usual. It is worth while going to the additional expense of a $2\frac{1}{2}$-in. tyred rear wheel as compared with the more usual $2\frac{1}{4}$-in., as this permits of a stouter tyre being used and one better calculated to withstand the extra strain thrown upon it, the only drawback to the practice of employing different sized tyres being that interchangeability is sacrificed. The side car may not be a mechanically correct form of attachment, but it is an uncommonly handy and pleasant one to use, nevertheless.

The Cost of Motor Cycling.

It is generally supposed—by those who have not tried it—that motor cycling is an expensive pastime, and, inasmuch as

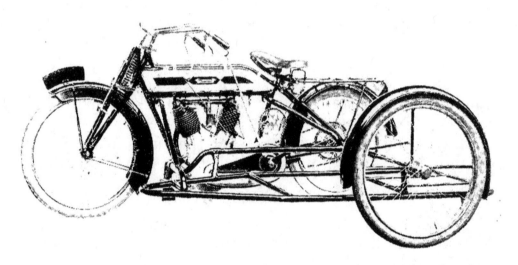

Fig. 204.—The "Quadrant" 7–9 h.-p. Motor Cycle and Side Car Chassis.

the terms "costly" and "expensive" are very elastic ones, they can be made to mean almost anything. It is quite possible to obtain a very large measure of enjoyment from motor cycling in return for a comparatively small outlay. The purchase of the machine is the most serious item, new machines of recognised make costing from about £35 to close upon £80; but a fair average figure is £48. It depends largely upon the make and horse-power, and whether the machine be a single-geared, clutch, or variable-geared one. Accessories, licensing, and other fees, help to mount up the initial cost; but, once this has been got over, the owner may, if he go carefully and intelligently to work, run the machine very cheaply as compared with the great benefits he will derive from its use.

The thing is to keep the machine as much as possible out of the repairer's hands, but not to neglect consulting a competent repairer, when the necessity arises, for work which cannot be done properly at home. Petrol (at the time of writing) costs from 1s. 7d. to 1s. 9d. per gallon. The best lubricating oils can be had for 1s. 6d. to 1s. 8d. per sealed quart tin, and less *pro rata*, if bought in larger quantities. These two commodities, and especially the former, need replenishing more frequently than anything else, and constitute a leading item in the running expenses. From 50 to 100 miles can be obtained from a gallon of petrol on average roads, according to h.-p. of motor and the work to be done, but a

FIG. 205.—The "Bat" T.T. Model Motor Bicycle for Fast Touring.

quart of engine oil goes very much farther. The author has obtained increased mileage from a gallon of benzole at 1s. 5d., and cannot trace any disadvantages due to the use of this slightly heavier spirit.

As before stated, good second-hand motor bicycles can be purchased for quite moderate sums, £20 securing a very fair magneto model, while for £30 (in the middle of the "off" season) some rare bargains can be obtained. The author commenced his motor cycling career some years ago on a machine costing, second-hand, £16, and it proved an unqualified success. In many cases motor bicycles have been purchased for as little as £8, and even then have proved satisfactory bargains. When the figure reaches such a low level as that, however, too much must not be looked for. The machine will necessarily be of old pattern, with accumulator ignition and some obsolete fittings, and may require a fair amount of work done upon it before it can be classed as in really satisfactory " going " order. If the purchaser is possessed of a workshop with a fair equipment of tools, he may easily be able to turn a poorly-conditioned second-hand motor bicycle, purchased at a low figure, into a very serviceable mount by his own energy and skill, combined with a moderate cash outlay. Passenger motor cycling naturally entails greater expense than otherwise would be incurred.

CHAPTER XIII.

CYCLE CARS, TYPES OF CARS, TRANSMISSION AND OTHER DETAILS.

SINCE the first edition of this Handbook was written a new style of light motor vehicle, incorporating in its design many essential features of the motor bicycle, has sprung into popular favour, and present indications warrant the assumption that this popularity will rapidly increase. These vehicles, some of which run upon three and others upon four wheels, may best be described as a compromise between the motor cycle with side car and the motor car proper. With the exception of the " Auto-Carrier " (A.C.) three-wheeler, the pioneer of its class, these light runabouts were first of all designed to carry one person only and for this reason were termed Monocars. Then, in response to the demand for a two-seater vehicle on the same lines, wider bodies, capable of seating two persons side-by-side, were introduced and the name Duocar applied. A type in which the driver and passenger are seated tandem-wise was also introduced into this country from France, and has since been taken up as a standard model by certain manufacturers in this country, and eventually, as the move-ment developed, a term which might be suitably employed to cover all classes of machines coming within this category became necessary, and after various suggestions of a more or less suitable nature had been put forward, the term " Cycle Car " was invented by Major Lindsay Lloyd, of the R.A.C., and this has now been generally adopted, and is recognised by all concerned.

Some of the so-called cycle cars are nothing more or less than motor cars on a smaller scale than usual, their design following the generally accepted lines of car practice ; but others are strictly what their name implies, namely—a motor cycle with a car body, and an additional wheel, or wheels ; one remove, and in most respects an advance, upon even the best of side-car combinations.

Of the various types of transmission employed in cycle

car design that in which belts are used is the most simple and direct, and which approximates most nearly to motor

FIG. 206.—The Author ("Phoenix"), with his 8 h.-p. Tandem Cycle Car.

cycle practice. Used in conjunction with a clutch and variable speed device, and with an air-cooled engine as the power-unit, the motor cyclist has at his disposal a method of propulsion, the features of which he has always been accustomed to, and the management of which entails no new difficulties when later he becomes a cycle car-ist. Then again, a very large class to whom the motor cycle has never appealed—while the motor car has always been beyond their means—are being attracted by these handy little vehicles, and many are taking up motoring who would never have done so under previous conditions.

The types of transmission adapted to cycle cars, other than belt drive, include friction wheels, cardan shaft and bevel gears, chains, with gear boxes, live axles, and so on, and various combinations of these systems. An objection to the belt system as at present applied is the lack of a " reverse," but this is practically the only point open to criticism in a really well-designed belt transmission. The cars, being light and handy, are easily turned, and although more room is needed, it requires little, if any, more physical effort to do this than is called for with a heavy side-car combination. Lack of the reverse is not confined to belt-driven cars only, and, after all, those who require maximum simplicity and cheap maintenance must be content to do without this one convenience in face of the advantages they secure in other directions. There are many ways of incorporating a reverse gear with belt drive, and apart from the question of added mechanism and cost there is no difficulty which could not be easily overcome in this direction.

The author has driven a cycle car equipped with belt transmission over three thousand miles in many parts of the country, and as a result of his experience with it is very well satisfied indeed.

The drive from the engine is by silent chain to countershaft, and from the pulleys on the latter to the rear wheels by Whittle belts, a change-speed gear of the expanding pulley and sliding back axle being fitted. Fig. 206 is a view of the machine. In the Rudge cycle car the drive is by belt throughout, instead of a chain being employed from engine to countershaft. The engine is a single-cylinder one of 5–6 h.-p. It is known as the " New Rudge " 750 c.c. model, and has a cylinder measuring 85 mm. bore by 135 mm. stroke. The engine drives the countershaft (on each end of which is mounted a pulley 6 ins. diameter) by means of a 1-in. rubber belt. On the counter-

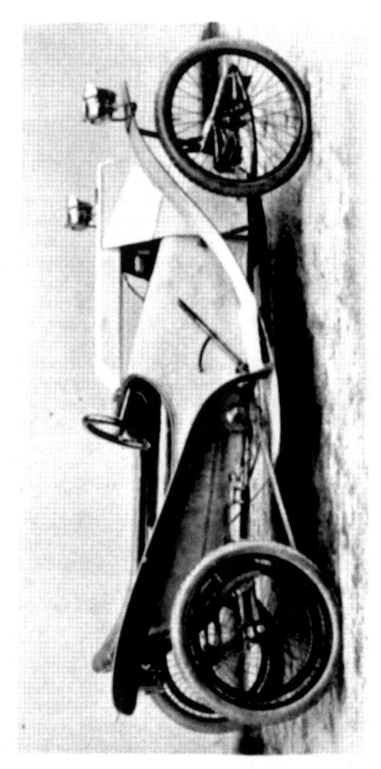

FIG. 207.—" Rudge " Cycle Car.

shaft, somewhere near its centre, is mounted a variable pulley, with clutch, the diameter of the pulley being $10\frac{1}{2}$ ins. The engine shaft carries a 6-in. pulley, built with separate flanges, each supported by a ball race, on similar lines to the well-known "Rudge-Multi," but without clutch. By the double reduction afforded, and the adoption of adjustable side pulleys on the countershaft, a range of gears may be obtained from $3\frac{1}{2}$ to 1 on the high, to 14 to 1 on the low, and ten changes of speed are rendered possible. The belt tension is uniform, except at starting, which latter operation is effected by means of a gear lever inside the car, which is pushed forward for high gear and pulled back for low gears. In traffic the car is controlled on the clutch.

Fig. 208.—The "Humberette" Cycle Car, with Air-cooled Engine, Cardan Shaft and Bevel Gears, Three Speeds, Forward and Reverse.

High-tension magneto ignition is provided, and the carburettor is of the latest Senspray pattern. The petrol tank is fitted above the engine, and holds four gallons. It is, in reality, divided into two separate tanks each with its own petrol tap. Both tanks can, however, be filled up at the same time.

The body of this car is of somewhat unconventional design. It is carried very low and the whole turnout presents a speedy and taking appearance. Trial of the machine on the road soon demonstrates that the first characteristic is not confined to looks alone, as the car is very fast and a good hill-climber.

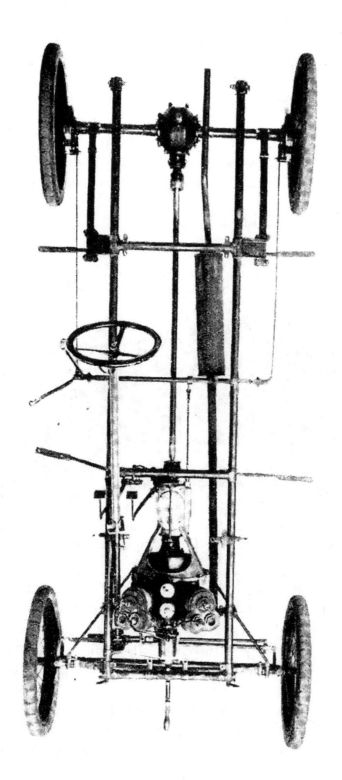

FIG. 208A.— Chassis, Engine, and Transmission of "Humberette" Cycle Car.

It remains to be seen, however, whether, with continued service, the two short belts of the final drive (from counter-shaft to rear wheels) will not show rapid deterioration. The large diameter of the side pulleys will to some extent mitigate this, and the fact of there being two belts instead of one, as in a motor cycle, is also in favour of the cycle car.

FIG. 208B.—Engine of the "Humberette" Cycle Car.

The presence of a second belt on cycle cars has some-times led to a misadventure, owing to the fact that the car continues its progress unchecked, even though one of the belts falls from the pulleys and is left lying unnoticed on the road behind. This has happened not infrequently owing to fasteners breaking, coming unfastened, or pulling out from the belt. The driver has proceeded in ignorance of what has happened, and only becomes aware that he has lost the belt on looking round the machine after arrival at home. With the Whittle belts, which have no fastener and are enormously strong, this contingency is almost, if not quite, impossible, and it is for this reason partly that the belt named has been standardised on certain cycle cars, having this type of transmission.

Of the cycle cars in which a more positive class of transmission is employed, the "Humberette" occupies a position

Fig. 208c.—The "Crescent' Friction-driven Cycle Car.

in the front rank. It is a sociable seater, propelled by a V-type air-cooled twin-cylinder engine, 84 mm. bore by 90 mm. stroke. The engine is set transversely to the chassis or framing, and in a position at the front of the car with the valve pockets forward, where they obtain the maximum cooling effect while the valves themselves are readily accessible. Transmission is by a leather-faced cone clutch with bevel gear driven by a propeller shaft from a gear-box to the live rear axle. There are three forward speeds and a reverse, and the engine is controlled by ignition and throttle levers placed on the steering column. There is also a foot accelerator for operating the throttle independently, although in conjunction with the hand lever. The springing of this little car is excellent, and the body provides comfortable seating for two persons side by side. The finish is of the high class usually associated with the firm of Humber, Ltd., and the design in all its details shows concentration of thought and the most careful workmanship.

FIG. 208D.—The "Crescent" Friction-driven Cycle Car. Twin air-cooled engine set transversely at front of chassis actuating propeller shaft and friction wheels.

The "G.W.K." is a friction driven cycle car, and is one of the most successful vehicles of its kind yet introduced. It is sometimes asserted that friction drive

18

entails certain well-defined disadvantages, and certainly where this transmission is employed for vehicles of great weight and high engine power this is true. For light cars of the runabout type, however, it has proved entirely satisfactory although, with it, rather more practice is required

FIG. 208E.—The "Crescent" Friction-driven Cycle Car. Rear view of Chassis showing friction discs. Chain drive to rear axle.

at first in the changing of gears than with other types, if the best results are to be obtained.

Friction wheels provide, as a matter of fact, one of the most simple possible forms of varying gear ratios, but there is a certain knack in handling the control mechanism which

must be acquired, and faulty manipulation sometimes leads
to the formation of flats on the discs, which latter then have
to be renewed. The proper handling of the mechanism is a
thing easily acquired, however, and practice has clearly
demonstrated the fact that the amount of horse-power required to propel vehicles of the cycle car type, even under
the most adverse possible circumstances, can be satisfactorily
transmitted by means of the friction principle, in spite of the
small area of the surfaces in contact for communicating the
power from the driving to the driven members, while the
system provides the simplest of all known reverse mechanism.

Another cycle car of the friction-drive type is the " Crescent," of which illustrations are given on pages 272-3-4. The
engine, an air-cooled twin, is placed athwart the chassis at the

FIG. 209.—The " Premier " Sociable Cycle Car.

front end, and the drive is by cardan shaft to the main or driving friction disc, and from the subsidiary shaft (on which is
mounted the driven friction wheel) by chain to the rear axle.
A neat and comfortable type of body is fitted, and the general
appearance of the car may be gathered from the illustration
(Fig. 208c).

Another handy little four-wheeled cycle car is the
" Premier," the engine of which is an air-cooled twin 85 mm.
by 88 mm., = 998 c.c. In this design the engine is carried
as on a motor cycle—that is, with the " run " of the machine,
and not transversely as in the " Humberette." The transmission is by chains throughout, i.e., from the engine shaft
sprocket to a two-speed gear-box, and thence by a second
chain to the rear axle differential mounted on the back axle,

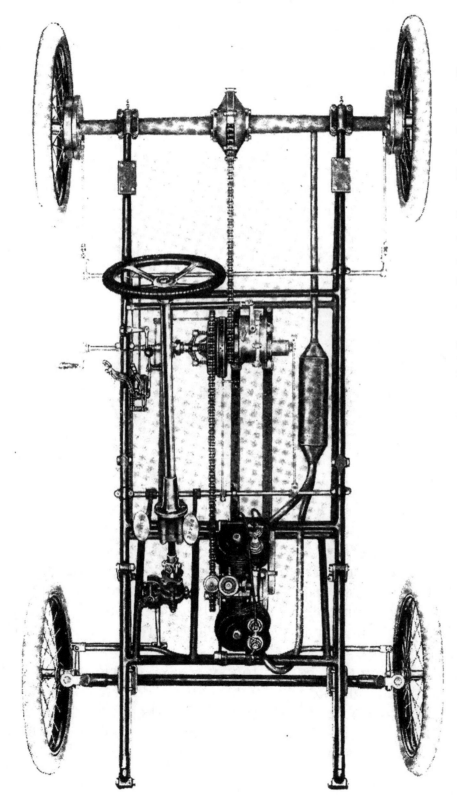

FIG. 209A.—Chassis, Engine, and Transmission of " Premier " Cycle Car, Chain Drive. Two Speeds and Reverse.

The plan view of the chassis, Fig. 209A, shows the engine, transmission, gearbox, etc., completely assembled, and this illustration will allow potential cycle car-ists and others to form a good idea of the arrangement of the various mechanisms which go to make up the construction of the working portion of one of many good examples of these little vehicles. The "Premier" car has wheel and segment steering, tangent wire wheels with heavy rims, and a comfortable two-seater body arranged on the sociable or side-by-side principle.

Of the three-wheeled cycle cars, the "A.C." and "Morgan" are the best known—the former, as already stated, claiming

FIG. 210.—The "A.C." (Auto Carrier) Three-wheeler, Single-cylinder (Air-cooled) and Chain Drive.

the credit of being first in the field. It has now been running successfully for several years, and is considered one of the best of the light motor runabouts at present on the market. It has a two-seater sociable type body, and is also made with a three-seated body capable of accommodating three adult persons. The engine in this design is placed aft between the rider's seat and the back wheel. It is a 5–6 h.-p. air-cooled unit, transmitting its power by means of a chain to the single back wheel in which is located a two-speed gear mechanism having incorporated with it a clutch of the multiple disc type. The two-speed gear is of the epicyclic pattern, with ratios $4\frac{1}{2}$ to 1 for the high and 9 to 1 for the low gear. Steering is effected by means of a tiller giving a very sensitive movement to the front wheels. The frame is built up of ash and steel combined, and is strong without being unduly weighty. It might be thought from a casual consideration of the design that the engine was unfavourably situated for cooling by air. Yet, as a matter of fact, overheating is an unheard of thing with this type, and although the engine is not one of the largest, it develops sufficient power to take the vehicle up the steepest of ordinary main road hills on the low gear without exhibiting signals of distress. Although not such a swift climber as some of the twin-cylinder cycle cars, nor quite so fast on the level

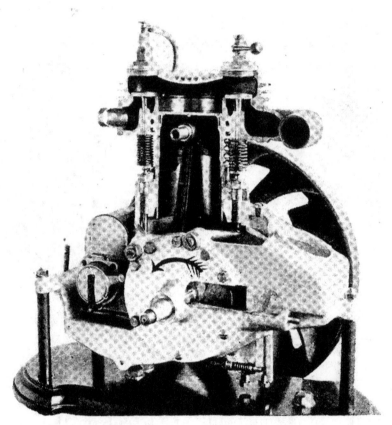

Fig. 210A.— View showing the "A.C." Single-cylinder Engine in Section

perhaps, the "A.C." in the hands of an ordinarily capable driver can achieve wonders, both as regards the surmounting of hills and fast travelling on the level. The makers (Auto Carriers, Ltd.) now manufacture four-wheeled cycle cars in addition to this three-wheeler. One type has a four cylinder water-cooled engine, while the other—a cheaper model—has the single cylinder type of engine shown in Fig. 210A.

The Morgan, Figs. 211 and 212, has a twin-cylinder air-cooled engine, placed athwart the car in front. It, also, is a very popular make and among other achievements it holds the Cycle Car Trophy—a prize offered for the highest number of miles to be covered by a cycle car in one hour on Brooklands track, during the year 1912.   The actual distance travelled by the Morgan in creating the record was 59 miles, 1,120 yards., The car has also scored consistently on the road in open competition.   In this design the transmission system consists of a cardan, or propeller shaft, actuating bevel gearing on the countershaft, whence the drive is taken up by chains to the back wheel.   A leather-lined cone clutch is used.

As a new introduction for the 1913 season the Matchless people (H. Collier & Sons, Ltd.) introduced a three-wheeled cycle-car similar in general design to the Morgan.

FIG. 211.—Front View of the " Morgan " Three-wheeled Cycle Car, with Twin Engine.

It has a 9 h.-p. twin-cylinder, air-cooled engine, with cardan shaft and bevel drive.   A two-speed gear box is used.   The body is, if anything, slightly more roomy than the other three-wheeled types, and it is carried higher above the ground. Side doors are fitted and also a hood and screen.

In addition to the above cited examples of cycle-car design there are numerous others incorporating the same or similar features and ranging in price from about £70 for a single-seater to as much as £185.   When the latter price is reached, however, we are getting more into the small car region and, to the author's way of thinking, rather beyond the scope of what can be legitimately termed a cycle-car. A limit has been fixed by the A.C.U. as regards the weight of vehicles eligible to be classed as cycle-cars ; and the engine capacity has likewise been defined. The weight of the chassis must not exceed 6 cwts. (or 7 cwts. where the body and frame are built as one), while the engine capacity permissible as a maximum is 1,100 c.c. Nothing has been done in the way of fixing a price limit, and it seems unlikely that such a step will be taken, although, as a means towards checking over-development, it would certainly have that effect.

The most simple type of cycle-car is that with a single-

FIG. 212.—Side View of the " Morgan " Cycle Car.

cylinder, air-cooled engine, and belt or chain drive, free-engine clutch and two-speed gear. For single seaters the engine need not be a large one, and indeed the whole vehicle, except in the matter of additional comfort, is but little more than a two-speed motor cycle, which does not need balancing. For the heavier two-seater types, however, a twin-cylinder engine of from 7 to 9 h.-p. is an advantage, and to many minds the engine should be of the water-cooled, instead of the air-cooled, pattern. The author does not share this opinion now, after some experience of both types, although, to be perfectly candid, before taking up the subject of cycle-cars from a practical standpoint, he was an advocate of water cooling for this particular purpose.

If the engine is placed where it is exposed to the full effect of the air currents, and is suitably geared and of sufficient power for the work it has to do, then, unless carelessly or ignorantly driven, it will not overheat any more than it does on a motor cycle. Water-cooling, after all, *is* a complication, and in adopting it the tendency is to break away again from motor cycle practice, and to follow in one more direction, that of the motor car. There is, maybe, no harm in this, but if the cycle-car is to remain a cycle-car and not become merged

Fig. 213.—" J.A.P." 10 h.-p. Air-cooled Engine for Cycle Cars, Cylinders at 90 deg., Magneto and Carburettor between them.

into the small motor car proper, then the more like a motor cycle it is the better for its prospects.

One marked advantage of the cycle-car over the side car combination is that the driver shares with his passenger an additional degree of comfort. No longer is it necessary to dress up specially for every outing, for the donning of ordinary clothes is sufficient, and one arrives at the end of the journey in as respectable a condition as at the start. The flinging of oil and mud to which we have submitted more or less cheerfully on the motor cycle has no more to be endured, and these are weighty considerations with many, although the younger generation will probably scoff at them.

As regards the cost of running a four-wheeled cycle-car: the author has found that he can average 50 miles to the gallon under ordinary circumstances, *i.e.*, a not too hilly road or adverse weather conditions, and from 40 to 45 miles per gallon with circumstances against him, as, for instance, a difficult road, with a headwind or an unusually heavy passenger and luggage. Oil consumption approximates to that usual with the same h.-p. engine fitted to a motor cycle hauling a side car, *i.e.*, a quart for 200–250 miles, the average being, again, governed by circumstances. Most cycle-cars are fitted with drip-feed

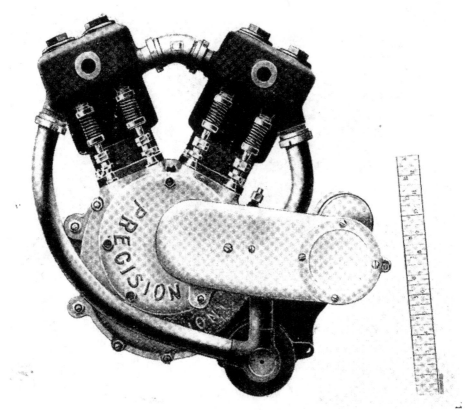

FIG. 214.—The " Precision " Water-cooled Cycle Car Engine, 8 h.-p.

lubricators and many, including the one used by the author, have a hand feed-pump as well.

The Inland Revenue taxes payable on cycle-cars range from £1 to as much as £3 3s. Any car coming within the weight limit of 3 cwts. unladen is rated as a motor cycle, with the £1 tax, irrespective of engine capacity. A few three-wheeled cycle-cars conform to this requirement, but lately the authorities have shown a tendency to more closely enquire into the weight of such vehicles. Four-wheeled cars, with engines rated up to 6½ h.-p. (according to the official definition of h.-p.), and weighing over 3 cwt., are liable for the £2 2s. tax, and above this h.-p. up to 12 h.-p. the tax is £3 3s. ; again, up to 3 cwts. the registration fee is 5s. (as before in accordance with motor cycle rating) ; above that weight £1 is chargeable, as in the case of motor cars.

In the author's experience, the wear of driving wheel tyres is less than on a powerful motor bicycle doing side-car work, this being due, no doubt, to the fact that, in the first place the wear is distributed over two wheels instead of being concentrated on one, and, secondly, to the absence of side strain, such as is set up by a side car.

Wheel steering is a tremendous improvement over handlebar steering. It is much more sensitive and affords greater control than does the other method. As regards cornering : with a four-wheeled cycle-car corners may be taken with impunity at speeds which would almost certainly lead to a spill with a side-car combination. This indeed is one of the strong points in favour of the cycle-car. The tandem type of car is far more sociable than those who have not tried it imagine. All the talking that is usually done while motoring may be indulged in without inconvenience with this pattern body, and owing to its decreased width less resistance is offered ; consequently the machine is fast and it holds the road very well indeed at high speeds. The four-wheeled types of car have the advantage that all wheels are immediately accessible for tyre repairs or renewals.

The garaging difficulty is a somewhat acute one to those living in large towns and this militates to some extent against the growth of the cycle-car movement. Even that obstacle, however, appears to have little effect upon the progress of the movement as a whole.

CHAPTER XIV.

ACCESSORIES AND EQUIPMENT.

THE experienced motor cyclist knows full well how necessary it is, if complete satisfaction is to be obtained from his machine, to be careful in the selection of the accessories he intends to use with it. As in the case of the machines themselves, there are different grades of quality in the various fitments made and sold for attachment to motor cycles, and the purchaser will do well to study the design of these and investigate the more important points connected with their manufacture.

There is a tendency among some motor cyclists to disregard the importance of this part of our subject and to cut down as much as they possibly can the amount spent on the accessories and equipment of their machines. This, to say the least of it, is a " penny wise and pound foolish " policy, for, unless the articles are thoroughly well-made and of good materials, they not only fail to give satisfaction while they last, but the " lasting " itself is only quite a brief business.

Motor Cycle Tyres.

First of all, the owner of a motor cycle should make very sure that the *tyres* fitted to his mount are good ones, and of adequate proportions for the work to be done ; and if he takes the following short table as his guide he cannot go far wrong :—

For motor bicycles of $1\frac{1}{2}$ h.-p. to 2 h.-p., $1\frac{3}{4}$-in. tyres.

For motor bicycles over 2 h.-p., up to 3 h.-p., 2-in. tyres.

For motor bicycles of $3\frac{1}{2}$ h.-p. to 4 h.-p., $2\frac{1}{4}$-in. tyres.

For motor bicycles of 5 h.-p. to 8 h.-p., $2\frac{1}{2}$-in. or 650 by 65 tyres.

For the rear wheels of tricars and special machines for carrying two passengers (apart from bicycle and side-car combinations), a " Voiturette " tyre (about 3-in.) will give the best results.

In order to prevent, as far as possible, the danger of side-slipping, motor cycle tyres are now-a-days made with rubber studs or other projections on the treads, and in some cases the non-slipping qualities of these are so small as to be almost negligible. Others, however, are perfectly efficient, and, speaking from his own experience, the author can testify to the good qualities of the Kempshall tyre, the hard-wearing qualities and general efficiency of which are now so universally recognised as to need no recapitulation here. The author

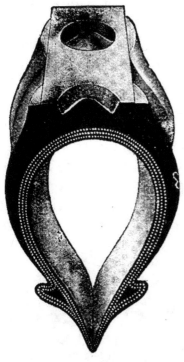

Fig. 215.—Section of "Kempshall" Motor Cycle Tyre Outer Cover.

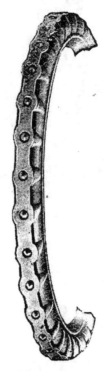

Fig. 215A. The "Kempshall" non-skid Tyre.

has always experienced the greatest satisfaction from these tyres, and can cordially recommend them. Another tyre which has given him good service is the Severn, a cheaper but hard-wearing product. Other excellent productions are the Dunlop and that known as the "Palmer Cord" tyre.

Considerable popularity is enjoyed by the detachable type of inner tube for motor cycle use. This may be either open- or butt-ended; but, in any case, its removal from the wheel when the rider requires either to repair a puncture or to replace the tube by another is greatly facilitated, as, once the outer cover has been released from one side of the rim, the tube may be taken right away in a few seconds and another put in its place.

FIG. 216.—The "Palmer Cord" Motor Cycle Tyre.

The butt-ended tube is, perhaps, the easier to fix in place and inflate, as in its case there is no joint to trouble about, whereas in the open-ended detachable tube, unless the jointure between the two ends is carefully made, so as to be absolutely air-tight, it becomes impossible to inflate the tube at all. It is, however, a comparatively easy job to secure this air-tight joint between the two ends. It is advisable, when entering the ends of the tube in the butt-ended type to sprinkle them well with French chalk, so as to prevent friction and the probability of rubbing through of the tube.

The use of steel studs in the tyres of motor cycles has gone somewhat out of favour of late years, although one very good tyre has rubber diagonal bars with steel studs in

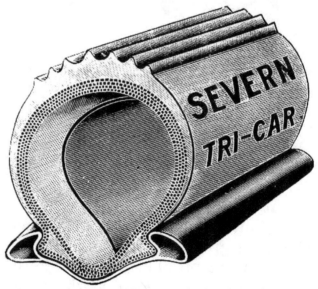

FIG. 216A.—The "Severn" "Tri-car" Tyre.

between. Rubber studs are very effective as skid preventers, and they last a reasonable time ; they hold the road well in very loose dust and on frozen roads, whereas steel-studded treads do neither. As a general rule, motor cycle tyres are

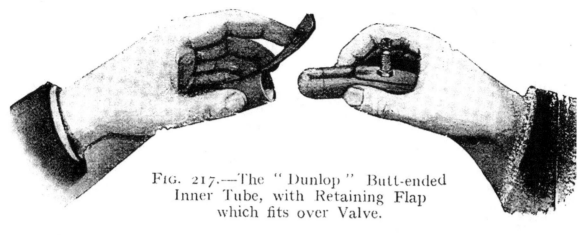

FIG. 217.—The "Dunlop" Butt-ended
Inner Tube, with Retaining Flap
which fits over Valve.

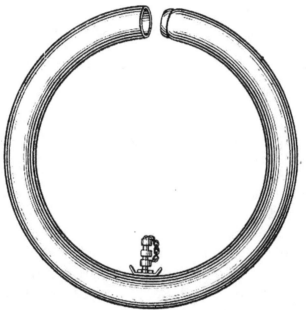

FIG. 218.—The "Rich" Detachable Inner Tube (open-ended).

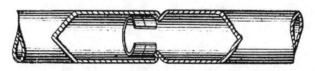

FIG. 219.—How the Air-tight Joint is made in the "Rich" Inner Tube.

made with beaded edges instead of being wired on. It is advisable to keep them pumped up hard while riding and always to carry a *really* good tyre-repairing outfit and suitably proportioned levers for removing the outer cover to repair or exchange a tube. A decent sized and well-made pump is also a fixed necessity.

By far the greater proportion of the wear falls upon the rear (driving) wheel tyre, this having to withstand greater weight and added friction ; the power of the engine in driving the machine being, of course, transmitted to the road through the rear wheel. It is a good plan to change the tyres over when the non-skid tread on the back one is worn down, this tyre being placed on the front wheel and that of the latter going on to the rear wheel, this, of course, being rendered impossible in the case of unequal sizes of tyres. When both are quite smooth—but not seriously cut or otherwise damaged —a new tread can be vulcanised on at a cost of from 14s. 6d. to one guinea per tyre, and in this way considerable further service may be obtained.

If matters are allowed to go too far, however, and the inner fabric of the cover is cut about much, it may be found impossible to carry out the vulcanising process with success ; and in any case the inner portion of the cover would have to be specially repaired, thus adding to the cost of the work. The strength of the walls of the tyre is an important consideration, for the addition of a heavy vulcanised tread increases the strain thrown upon them.

If good tyres are necessary on a motor bicycle so is a good belt, in order that the power developed by the engine may be transmitted with efficiency, and as large a percentage of it as possible utilised in driving the machine (in accordance with the output at various speeds), for otherwise there cannot possibly be satisfactory working.

Belts and Belt Fasteners.

As remarked in a previous chapter, the belt must possess strength, flexiblity, and a sure grip of the pulleys ; and if only one of these attributes is missing, the belt, as a whole, will fail to come up to what is required of it. Rubber and canvas belts are very popular, being clean to handle and capable of more minute adjustment in respect of length than are most leather belts. The latter, however, as before said, last longer and seldom break ; they are very strong, but need carefully attending to ; occasional lubrication to keep them pliable, and somewhat frequent scraping of the effective surfaces on which, in course of working, grit and grease accumulate, being required.

The two leather belts specially referred to in a previous chapter will both be found highly suitable for their purpose, and both have the advantage of being easily shortened when required.

In selecting a belt-fastener, make sure that it is easily detachable, for many so-called " detachable " fasteners require a lot of patience and pulling about to get them apart. Motor cyclists are not perhaps the most patient of all people, they being accustomed to move along quickly, and to be kept hanging about trying to get the belt fastener adrift or re-assembled, perhaps in the rain or intense cold, suggests profanity with not a few.

In the author's opinion the best type of belt fastener is that which, in addition to being detachable, allows of variation in its length by using longer or shorter links or hooks as required, and the manner in which this is effected may be gathered from the sketches appearing in Chapter VIII. The fastener, to be satisfactory, must be strongly made, and care should be taken to use the proper size of fastener as made for attachment to a given width of belt.

Tyres and belts may be better classed as " equipment " than as " accessories " ; but they have been purposely included in the present chapter as coming within the range of renewable parts, in the purchase of which the motor cyclist will have to exercise his own judgment.

The Lamp Question.

A very essential accessory on a motor cycle, and one which should *always* be of good quality, is the head-lamp. It is quite possible now-a-days to buy a large, showy-looking lamp for quite a moderate sum, but it is false economy to do so. The lamp should, above all, be substantially made, possess a first-class lens, and have riveted instead of brazed or soldered joints.

It is not always the largest and heaviest lamps which give the greatest amount of light, neither should the intending purchaser allow himself to be carried away too far by the

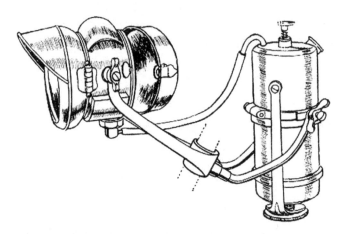

FIG. 220.—The "F.R.S." Acetylene Headlight for Motor Cycles.

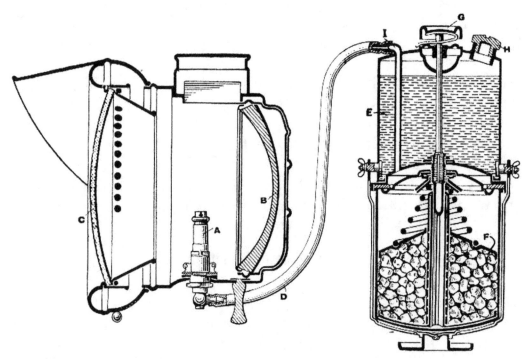

FIG. 221.—Sectional Drawing of Separate Generator Type
Acetylene Headlight.

(A)—Burner.   (B)—Lens Mirror.   (C)—Front Lens.   (D)—Rubber
Gas Tubing.   (E)—Water Compartment of Generator.   (F)—Carbide
Container of Generator.   (G)—Needle Valve Adjuster   (H)—Water
Filler Cap.   (I)—Gas Pipe.

mere outward shape and appearance of the lamp. Space prohibits anything in the nature of even a brief treatise on the subject of light-giving in connection with motor cycle lamps; but it may be said that a really good head-light will throw a beam of 600 ft. to 800 ft. in front of the machine, and nothing is more strictly necessary, where much night riding on main or other roads in the country has to be done, than that a powerful lamp should be used.

Most motor cycle lamps are of the acetylene gas burning type, and most are provided with separate generators, so that a powerful light may be obtained, and the necessity of

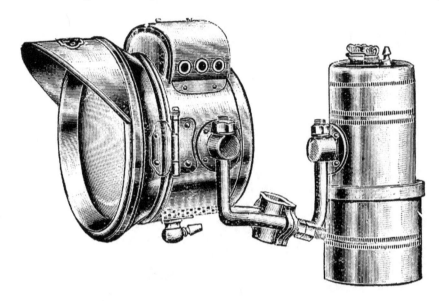

FIG. 222.—The " Service " P. & H. Acetylene Headlight.

recharging made as infrequent as possible without concentrating all the weight of the lamp in one unit. Smaller lamps having the generator in one with them and of rather less capacity are made, these being known as of the " self-contained " type, and they have the advantage of clipping all in one piece on to the lamp bracket.

In some few cases the makers have designed separate generator lamps, in which the generator is attached to the same bracket as the lamp itself. In this case the clip is fixed, as a rule, to the handle-bar stem, as seen in the accompanying illustration (Fig 222), which shows the " Service P. & H." headlight—a very powerful light-giver, substantially made, and, indeed, a high-class production all through.

The F.R.S. lamp is also illustrated. This is rather larger

than others, and the light it gives is exceedingly strong and
penetrating. The lamp is scientifically designed and well
made. The makers claim that it gives the world's record
beam of 1,000 feet ; a generator of the " riddling " type is provided. Other well-known motor cycle lamps are the Lucas
(" King of the Road ") and " Seabrook Solar," and the
reader will be perfectly safe in investing in either.

It is highly necessary to employ a very strong clip for
attaching the lamp and generator to the machine. Makers
are fully alive to this fact *now*, but a few years ago it was a
common thing for motor cyclists to have their lamps smashed,
or at least seriously knocked about, as a result of the clip
breaking and the lamp being dashed to the ground.

Satisfactory results can only be obtained, even from the

FIG. 223.—The " Glare " Electric Headlight.

very best lamp made, so long as the user is prepared to keep
the generator clean internally, the burner free from getting
sooted up, and only a really good brand of carbide should
be used.

Electric headlights are sometimes used on motor cycles,
and the author has experienced much satisfaction from
employing the " Glare," made by Messrs. Greenwood & Co.,
of Halifax. This lamp is provided with a hinge, so that it
can be tilted up to read the directions on sign posts, and the
current is supplied by an accumulator neatly enclosed in a
metal case, lined with india-rubber, and provided with clips
for attachment to the frame or handle-bar of the machine.
These lamps are very useful indeed for inspection purposes,
as with them there is no fear of an explosion should there

be an overflow of petrol. The lamp question is an important one with the motor cyclist, and he is urged not to grudge the money required to buy a really good article.

Horns, Sirens, and Exhaust Whistles.

We may next devote a little attention to the subject of appliances designed for creating audible warning of the motor cyclist's approach. These include horns (or hooters, as some prefer to call them), exhaust whistles, and sirens. The first-named are most largely used, but the other two, and especially the whistle, enjoy a fairly considerable vogue.

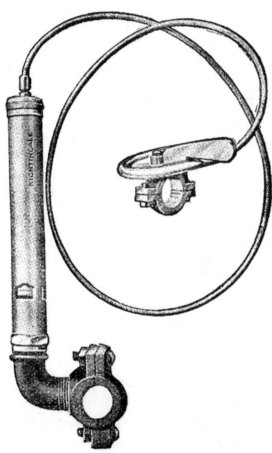

FIG. 224.—The "Nightingale" Exhaust Whistle, with Handle-bar Control.

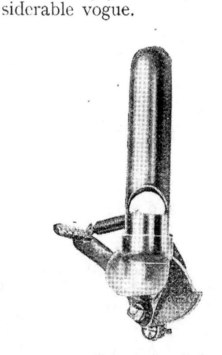

FIG. 225.—Single-note Exhaust Whistle, Foot-operated.

Again, it is false economy to buy any but a strongly and well-made horn, and special attention should be devoted by the purchaser to the clip attachment. This, in very many cases, is quite inadequate for its purpose, and the rider is subjected to the annoyance of its breaking, while a common fault is that the horn either revolves around the handle-bar or goes dancing about upon the latter, setting up an irritating chatter, and gradually wearing away the plating. A horn is of very little use unless :

(1) It creates a loud, sonorous sound ;
(2) Has a fair-sized rubber bulb, and well-made reed ; and
(3) A strong, properly fitting clip.

Tri-note or " fanfare " horns, having three separate horns attached to one stem and bulb, are sometimes used. In the author's view this type is not to be recommended. The noise it makes is an especially irritating one to other road-users, and it is necessarily a more complicated piece of construction than the single type of horn. A good, substantial, well-made single- or double-twist horn, leaves little to be desired.

Exhaust whistles are very effective as road clearers. They perhaps work better with a twin engine, owing to the more continuous exhaust ; but they also do very well on a single-cylinder engine, and command attention where the more usual " hooter " fails to do so. The " Nightingale " whistle, operated from the handlebar, produces a loud but musical variety of notes, and is certainly superior to the ordinary whistle with a fixed note, and is a strong and well-made article. The author has used this appliance with much success.

Motor Cycle Saddles.

The saddle is an all-important adjunct of the motor cycle, and stress has already been laid, in the chapter on " Springing," on the necessity for using one which is properly designed and made.

The saddle should be springy, without being what may perhaps be best described as " bouncing." The rider desires to be insulated, as far as possible, from vibration, but without being nearly jerked out of his seat. His comfort depends very largely indeed upon the quality and character of the saddle, and only the best-known makes should appeal to him.

The Brooks B-500 motor cycle saddle, with padded top and compound coiled springs, as illustrated, on page 201, is a very comfortable one indeed, and much of the success and high reputation rightly credited to it is due to the fact that the springs are proportioned to the particular rider's weight. This fact should be borne in mind when purchasing a saddle of this kind, or when the bicycle one has selected is fitted with it. The makers always desire to know the rider's weight, and they will then supply a saddle having springs suitable thereto.

A good saddle costs as much as from 23s. to 45s., and even higher, and it is cheaper in the long run to give a good price at the outset.

Luggage Carrier and Stand.

The purchaser of a motor cycle should carefully inspect

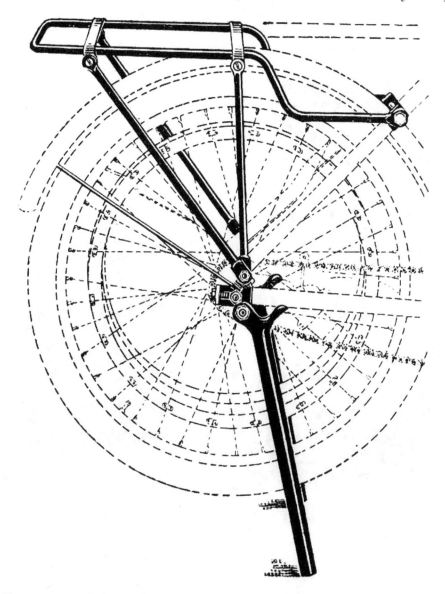

FIG. 226.—Luggage Carrier and Stand for Motor Cycles.

the stand and carrier supplied with it. These must be strongly made, and the stand (by means of which the back wheel is held off the ground) should have some easy means of fastening when not in use. In some cases the stands are made to spring upwards, and automatically lock themselves in position as the wheel is dropped to the ground, and if properly designed this is a great convenience.

It is common practice to make the carrier of tubular material, thus adding strength to the construction and reducing the chances of the bags or other packages carried upon them getting chafed. The practice, at an earlier

FIG. 227.—Combined Tube and Belt Case (Priory Accessories Company, Coventry).

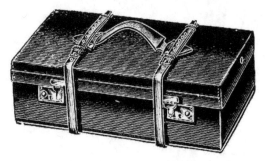

Fig. 228. Metal-lined Tool Bag.

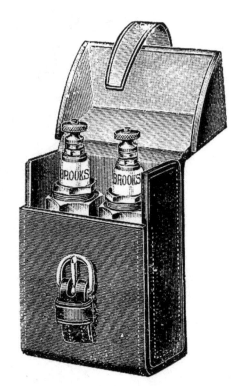

FIG. 227A.—Case for Carrying Two Sparking Plugs (Brooks).

stage, was to make the stand and carrier in one, and every time the rider wished to perch his machine up he had to remove the tool-bag, etc., before he could do so ; and, further, as the carrier was mounted on the rear wheel spindle, when he wanted to take the wheel out the stand became inoperative. Now-a-days, however, the two are made separately from one another, and, moreover, the legs of the stand are hinged on to a lug on the chain stays, so that in taking out the back wheel the stand may be used for supporting the machine.

Stands are made either with separately controlled legs or with a cross-piece connecting them, and, everything considered, the latter plan is perhaps the better one, except where a side car is used, for then only one leg—the outer one—is required. The makers usually supply the all-in-one-piece type of stand, it being more simple to make, and very easily attached. Front wheel stands are more generally fitted now than formerly, and they should be standardised as a regular feature of the equipment.

Tool Bags and Equipment.

The average motor cyclist, when setting out for a long ride, takes a spare inner tube and a spare belt with him.

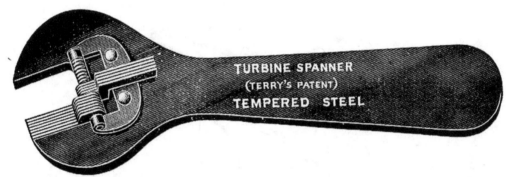

FIG. 229.—Terry's Turbine Spanner.

Sometimes—indeed, very often—he ties them up and hangs them on the machine, wherever he can find a place, by means of string or a strap. Much better is it to invest in one of the neat, combined tube and belt cases now on the market. These are strongly made of leather, and of circular shape. The tube is coiled up in a separate central compartment, while the belt space lies outside of this, as seen in the illustration. Both articles are kept dry, free from getting chafed, and always ready for immediate use.

The tool bag (or bags) should also be capacious and *any-thing* but flimsily made. They have to withstand a lot of hard wear, and are preferably fitted with metal linings to prevent holes being worn through them by contact with the tools. The latter, or the heavier ones among them, shou... be wrapped up in pieces of rag, and matters should be organised that delicate spares, such as sparking plugs, carburettor and magneto parts (if carried), small nuts and washers, and other small sundries be kept apart in a bag or in separate cases (as Fig. 227A) by themselves. Spare valves should also be wrapped up and kept from contact with heavier items, and so should insulating tape, high-tension wire, and, in fact, anything likely to get easily damaged or broken.

Some idea has already been afforded of what the tool kit should comprise, and it is, of course, quite impossible to enlarge on that part of the subject in any degree of completeness. Special spanners and other appliances are to be had by the hundred from the leading accessory dealers and others, some being excellent, but others—well, not so good.

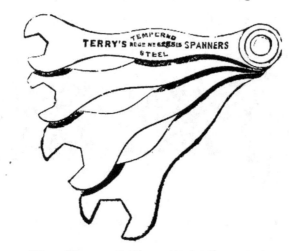

FIG. 230.—The Terry "Four-Fold" Folding Spanner.

The Terry spanners, shown on this and the next page, have been found extremely useful by the author, who makes a rule of always carrying them. They are strong without being bulky (a great point in their favour), and the adjustments permitted by the turbine spanner are very minutely graded.

A good and complete tyre repair outfit is compulsory, and the rider should never neglect to carry one with him. Never be deluded into buying cheap and nasty sparking plugs. Their use is fatal to good results. The same applies to belt fasteners, and, indeed, to every part—large and small—of the machine and its appurtenances.

Speedometers.

The use of a speed-indicating and mileage-recording

device on a motor cycle greatly adds, not only to the convenience, but also to the pleasure of the rider. It is extremely useful and interesting to know at what speed you

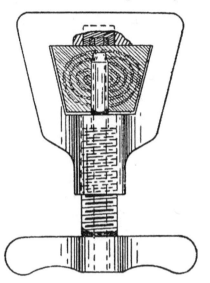

FIG. 232.—How the Belt Punch Works.

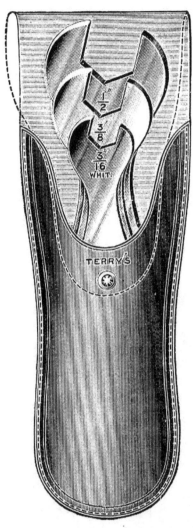

FIG. 231.—A useful case of Spanners (made by Terry and Sons, Redditch).

FIG. 233.—The "Sphinx" Double Belt Punch for two sizes of belts.

are running, and not only this, but it becomes almost imperative to be able to correctly judge the speed when travelling through ten-mile limits and wherever the police are known to be actively on the track of motorists. But if one is going to invest in a speedometer, and rely upon it as a means of gauging the speed of his machine, he must be very 'ware of the cheap inaccurate type of instrument which, so far from helping him to keep clear of a trap, will more likely than not lead him into one by showing a lower speed than that at which he is in reality travelling.

The "Watford" Speedometer.

The author has for some considerable time past been using a "Watford" speedometer made by Messrs. Nicole Nielsen and Co. Ltd., at their large and well-equipped works at Watford, Herts, a British production throughout. This instrument is of handsome appearance; the dial is very clearly

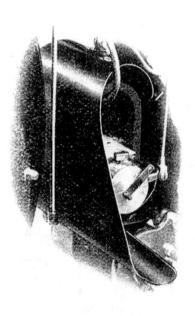

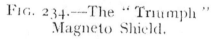

Fig. 234.—The "Triumph" Magneto Shield.

Fig. 235.—The "Watford" Motor Cycle Speedometer.

marked, and the accuracy of the indicating and recording mechanism is beyond criticism. A marked feature of the speedometer is the absolute steadiness of the pointer, even on the roughest roads and at the highest speeds. The author has seen these instruments in every stage of their construction at the Watford works, and can, therefore, testify to the accuracy and skill with which they are made, as well as to the excellence of their behaviour in actual use on the road.

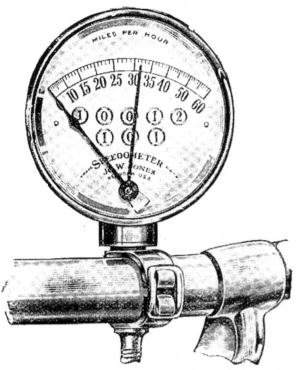

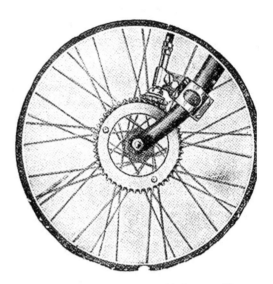

FIG. 235B.—The " Jones "
Speedometer (Front Wheel
Fittings).

FIG 235A.—The " Jones " Speedometer
Dial on Handle-bar.

FIG. 235C.—The new Gear-driven " Veeder " Cyclometer.

The instrument is made in several forms for motor cycle and cycle-car use ; the particular model in the author's possession being fitted with a trip recorder, registering each mile as it is completed, this being in addition, of course, to the indicator pointing up to 60 m.p.h. and the season's mileage recorder.

The Jones speedometer is another excellent example of this class of instrument. It is made in three forms for motor

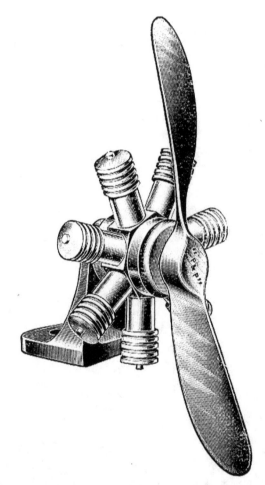

FIG. 236. Model Aeroplane Engine and Propeller. A " Mascot," sold by A. W. Gamage, Ltd.

cycle use, *i.e.*, (1) indicating up to 60 m.p.h. and showing the season's mileage ; (2) indicating to 60 m.p.h., showing season's and also daily mileage ; and (3) season or total mileage in tenths and miles, and also the daily mileage. This last-named model is fitted with maximum hand and instantaneous reset mechanism (Fig. 235A). The " Veeder," a cheaper but thoroughly reliable instrument for recording only the distance travelled by day and by year (without speed indicating), is made for attaching to the front wheel spindle. This instrument, as now constructed, is actuated by a gear or crown wheel affixed to the spokes of the wheel, as in the case of the speedometers, a much more accurate and satisfactory method than using a star wheel, as is usually done in devices of this kind.

Both appliances (Figs. 235A–235C) are supplied by Messrs. Markt & Co. (London), Ltd., 98-100, Clerkenwell Road, E.C.

Miscellaneous Accessories.

The carrying of a watch on the handle-bar tends to increase the pleasure, and is also a useful practice; special carriers are now sold, together with a cheap but good pattern watch, all ready to go straight on to the handle-bar.

Of " special " motor cycle accessories there are no end. Many of them can well be done without, except, perhaps, by the very fastidious. Remember that everything you add to the machine increases the load the engine has to propel, and it is astonishing how these accessories help to total up in the aggregate. Good and well-made fitments, of not too bulky or ponderous proportions, should be selected, and, lastly, don't have anything to do with anything smacking of the " freak " description. There is no room for these on your motor bicycle.

CHAPTER XV.

The Law in Relation to Motor Cycling.

The use of a motor vehicle on the public roads is permitted by law under certain conditions which must be strictly adhered to, and the motor cyclist is just as much bound to observe these conditions as is the owner of a 60 h.-p. motor-car.

Driving Licence.

He can purchase as many motor cycles as he likes, but, strictly speaking, he cannot even push one of them, if completely equipped for riding, along a public thoroughfare unless it carries identification numbers, while before it is possible (in a legal sense) to drive the machine a single yard a driving licence must be obtained. The intending purchaser of a motor cycle should apply to the County Council of his own or some other county for (1) a driving licence, and (2) registration numbers. These will each cost 5s., and before either is granted he has to fill up a form giving particulars of the machine, and his own name and address, say whether he has held a licence previously, and give various other particulars, all of which are, however, very quickly answered. If the applicant is under fourteen years of age, he will have to bide his time until he has reached that age.

Number Plates.

Having secured the driving licence and identification numbers, the latter must be painted on two separate plates — one for attachment to the front of the machine, and the other to the rear. The letters must be a distinct white on a black ground, each figure or letter measuring $1\frac{3}{4}$ ins. high by $1\frac{1}{4}$ ins. wide by 5-16ths in. thick everywhere. The distance between the top and bottom edges of the plate and those of the lettering must be $\frac{1}{4}$ in., while $\frac{1}{2}$ in. of margin must be left at each side.

The space between the letters and figures must be $\frac{3}{8}$ in.

The letter " I " and the number " 1 " need only be 5-16ths-in. wide.

Inland Revenue Licence.

In addition to the fees already mentioned it is necessary to take out an Inland Revenue, or " carriage " licence. This costs £1, and can be obtained at any post office. It requires annual renewal (during January) and lapses on December 31st in each year. Motor cyclists applying for carriage licences any time during the last quarter of the year receive an abatement of half the amount charged for the full year. The driving licence is also renewable annually.

The charges apply equally to a small-powered, light-weight motor cycle and to a high-powered passenger machine. Side cars (attached to motor cycles), tri-cars, and other three-wheeled motor vehicles weighing not more than 3 cwts. unloaded, at present come within the same scale ; attempts have been made in certain districts to levy a charge on side cars, but so far without success, but if a trailer or other type of passenger or luggage carrying attachment on two wheels is used, a fee of £2 2s. becomes payable. The charges in respect of cycle cars are given at the end of Chapter XIII.

Regulations Governing the Sale of Motor Cycles.

If the owner of a motor cycle sells his machine he will, if he desires to keep his original numbers for use on a new one, have to apply for their cancellation and re-issue to himself, and the fee for this is 5s. If, however, he sells the machine, and the new owner takes the numbers over with it, the latter can apply to the County Council for transference of the numbers to himself, and for this he will be required to pay 1s.

It is very necessary that *both* the old and the new owners should notify the authorities, and failure on the part of either to do so may result in a visit from a police officer with, possibly, more or less unpleasant consequences to follow.

Illumination of Number Plates.

One or other of the number plates on a motor bicycle must be illuminated at night, and the easiest way out of this is, of course, to place the front plate where the headlight can shed a part of its rays upon it. No need then to illuminate the rear number plate. The law does not require you to do so. With a cycle car the case is different, for then the rule which applies to motor cars is enforced.

Neither are you obliged to carry a number plate on the back of a side car, but if a trailer is used it must carry a plate. The best plan is to employ a number plate for the front wheel which is painted on both sides, and secured to the mudguard in a longitudinal position, so that the light from the lamp may play on both sides of it.

As regards a rear light, although there is no obligation to do so, it is well to carry one, and it should, of course, show a red warning to overtaking traffic.

Speed Regulations.

A maximum speed of 20 m.p.h. is permitted on the highways, and anyone found to be exceeding this may be stopped, his name and address taken, and a summons for " exceeding the limit " issued. He is liable to be fined for riding his machine while not having his licence upon him, and must be prepared to exhibit the licence whenever required to do so by a police officer.

Most towns (and very many villages) have a ten-mile speed limit—a few have a five-mile limit—and any excess of this may lead to an appearance in court. Other offences include riding to the common danger, or in a manner dangerous to the public, and cases have been known in which motorists have been convicted and fined under this heading when their speed was *below* that of the prescribed local limit, the policeman having given it in evidence that, in his opinion, the number of people in the street, or other circumstances, made it dangerous to proceed even at that moderate speed.

As an instance of this, a friend of the author's was summoned and fined at a Petty Sessions for having " driven a motor cycle in a manner dangerous, etc., " through the main street of a little village, and the police-sergeant in charge of the case admitted, under cross-examination, that the speed did not exceed 7 to 8 m.p.h., the rider pulling up in one-and-a-half lengths of the machine when, quite unexpectedly, he was called upon to do so.

Should the rider be stopped for exceeding the speed limit alone he must either be warned by the police, at the time, of intention to prosecute, or such notice must reach him within three months of that date. If it does not, and he is subsequently summoned, he should urge the lack of notice in his defence.*

Should he be carrying a speedometer, and be keeping down the speed by its aid to less than that alleged by the policeman, he should direct the latter's attention to the fact, and should take steps to have the instrument tested by the makers ; and then, if summoned, call their representative to give testimony to the fact of its accuracy. It seldom happens that the speedometer (if a good one, and properly fitted) tells an untruth.

Brake and other Regulations.

A motor cycle is required by law to carry two separately controlled brakes. A clear light in the direction of travel must be shown from one hour after sunset to one hour before sunrise. Audible warning of approach must be given whenever necessary, the rider must stop his engine in response to the upheld hand of anyone in charge of a restive horse, and if the machine is left unattended on the roadway steps must be taken to prevent its being started by others.

The Storage of Petrol.

Petrol may be stored up to a maximum quantity of 60 gallons without a licence. In London 48 gallons only is allowed. It must be kept in metal cans and stored apart from a dwelling-house. The rules in this respect are only applied with stringency where larger quantities of petrol are stocked. The keeping of a sealed can or two immediately adjacent to, or actually within, the house would hardly be objected to, but the motor cyclist who is wise generally keeps the spirit well away from all chance of its getting accidentally ignited.

---

* It does not do to rely on this point as it is frequently disallowed by country Benches.

A Final Word.

In conclusion, the author sincerely trusts that what has been written in the present volume may prove useful to its readers. An endeavour has been made to cover the ground (comprising the subject matter of the work) as thoroughly as possible within the space at command, but any suggestions which readers may be disposed to put forward as a means of improving future editions will receive his most careful consideration, and will indeed be very welcome.

HAMPTONS LTD., 12, 13 & 19, CURSITOR STREET, LONDON, E.C.

INDEX.

1913 TOURIST TROPHY RACE.

Mr. T. WOOD, riding a Scott two-stroke machine, fitted with $26 \times 2\frac{1}{4}$ in.

PALMER
CORD TYRES,

was FIRST in the SENIOR EVENT.

The BLUE RIBAND of the MOTOR CYCLING WORLD.

The triumph of the Palmer is increased by the fact that out of 84 starters only 4 machines were fitted with Palmer Cord Tyres, all of which were bought in the ordinary course, and no inducement whatever was offered either to the rider or manufacturer to fit them.

Mr. Wood used the same pair of tyres throughout the race, unlike other competitors he was not stopped once by tyre trouble.

Write for Motor Cycle Tyre Booklet.

THE PALMER TYRE, LTD.,
119, 121, 123, Shaftesbury Avenue, LONDON, W.C.

MOTOR CYCLE TYRE DEPOT:
103, ST. JOHN STREET, CLERKENWELL, E.C.

B & B
Carburetters

For Speed, Reliability, and Economy.

SPEED.

B. & B. Carburetters hold the Premier **World's Record** of the day—67 miles 782 yards in one hour on a 3½ h.-p. 499 c.c. Engine. This was done on an **ABSOLUTELY STANDARD CARBURETTER.**

RELIABILITY.

Mr. T. J. WATSON (an Amateur) WON the Special Sidecar Prize in the M.C.C. LONDON—LAND'S END run at Easter, riding a B. and B. on a 3½ h.-p. Swift, beating machines up to 6 h.-p. Twins!

ECONOMY,

On April 24th, 1913, Mr. JAMES, on a 6 h.-p. Zenith and Sidecar, in the OXFORD M.C.C. PETROL CONSUMPTION TRIAL, did 115·2 miles to the gallon. Mr. HILL, on a 2¼ Calcott, did 240 miles to the gallon, in the same trial in 1912, with a total weight of 315 lbs.

THE 1913 MODEL

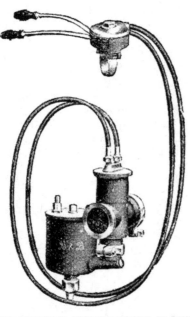

STRAIGHT-THROUGH PATTERN.

THE 1913

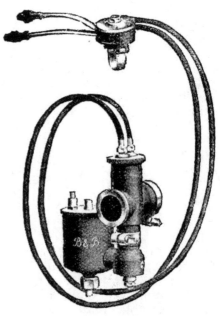

SINGLE-JET MODEL.

Write for Booklet, which includes "Hints and Tips" on Tuning-up. Free of Charge.

BROWN & BARLOW, Ltd.,
Westwood Road, Witton, BIRMINGHAM.

Lathe Aid

In Motor Cycle Repairs Means
SPEED — SATISFACTION — ECONOMY.

In fact, there is no better way of dealing with motoring repairs than the
" home-workshop " way. The installation of a lathe in the garage enables
all running repairs and a great many other kinds (those due to accidents,
etc.) to be done without delay ; it enables these repairs to be done the way
you want them done, and it certainly proves economical from the fact that
there is no subsequent repairer's bill to be faced.

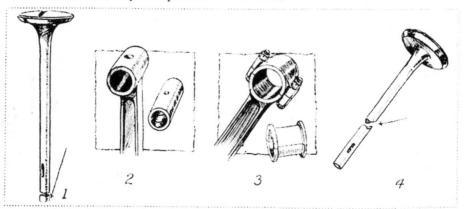

Just by way of showing what can be done, we give here four simple
examples of repairs that you might be faced with any day :—

No. 1.—Makers send out replacement valves a trifle longer than original
valves to allow for possible wear in cams and tappets. Often the allowance
is greater than necessary, in which case you will be faced with the job of
filing off, say, an ⅛ in. of nickel steel. On a lathe the job would be done in
a few seconds, either between centres or in a chuck ; it would be neat and
accurate, satisfactory to you and your machine.

No. 2.—A small end may require re-bushing : the knock is distinct and
demands attention. The dread of delays, once the engine is parted with,
will probably cause reluctance to have the job done. If you had a lathe
in your garage your first spare hour would see the repair completed and
the engine back in its frame, the turning down and scraping in of a new
bushing occupying only a few minutes.

No. 3.—Very much the same as No. 2.

No. 4.—The valve stem has broken, and you may find your spare
anything but a fit. It will be a *glorious* job grinding it gas-tight, but on a
lathe it would be a mere matter of seconds to turn the valve down and
grind it in.

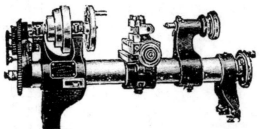

4-in. Centre Screw-
cutting, Boring and
Milling Lathe, the
tool for Motor Cy-
cle Repairs.

Write for full
particulars of this
lathe ; sent free
and post free.

DRUMMOND BROS., LTD.
INDVIEW WORKS — GUILDFORD — SURREY

BRITISH MADE

Watford
Speedometers

THE FIRST MOTOR CYCLE SPEEDOMETER TO ENTER AND PASS AN OFFICIALLY OBSERVED

Auto Cycle Union 3,000 Miles Test.

> **Extract from a letter from Mr. C. B. Donkin, of Woodthorpe, King's Heath.**
>
> *August 30th, 1913.*
>
> On January 13th of this year I fitted one of your **£3 3s.** Speedometers to my 2¾ h.-p. A. J. S. Motor Cycle. The instrument now registers 10,287 miles and has given no trouble at all.

PRICES—(Including complete transmission parts).

Type 706 for Motor Cycles or 706A for Cyclecars, 3-in. dial £3 3s.
Indicates speeds to 60 miles per hour, with total mileage counter to 10,000 miles and repeats.

Type 702 for Motor Cycles or 702A for Cyclecars... ... £4 4s.
Indicates speeds to 60 miles per hour, with total mileage counter to 10,000 miles and repeats.
Quickly re-set trip counter to 100 miles.

Type 700 for Motor Cycles or 700A for Cyclecars ... ... £5 5s.
Indicates speeds to 60 miles per hour, with total mileage counter to 10,000 miles and repeats.
Quickly re-set trip counter to 100 miles.   Maximum speed hand.

One Guinea Extra for addition of Type 399 Watch.

Write for Catalogue to—

NICOLE, NIELSEN & CO., LTD.,

Inventors and Patentees of the Chronograph 1862: Split Seconds, 1871; and Speedometer, 1904.

THE WATFORD SPEEDOMETER WORKS.

London Showrooms—14, SOHO SQ., W.

Telephone—2833 Central.
Telegrams—Niconielco, London.

Why do experienced riders cling to KEMPSHALL MOTOR CYCLE TYRES?

BECAUSE their reliability and endurance has been proved in every classic test, including the Scottish & A.C.U. Trials, in which competitors using Kempshalls have gained **NUMEROUS GOLD MEDALS AND AWARDS FOR FIVE CONSECUTIVE YEARS.** You can be confident when riding on Kempshalls. They hold up on anything. Test them yourself.

THE KEMPSHALL TYRE CO. (of Europe), Ltd., 97, 98, Long Acre, W.C.
Telephone: No. 244 Gerrard (2 lines).　　　Telegrams: "Studless, London."
Birmingham—Reginald G. Priest, 71, Lionel Street.　Paris—John Sheen, 46, Rue St. Charles.　New Zealand—Goldingham & Beckett, Ltd., Palmerston, N., New Zealand. Scotland—Percival E. Pole, 27, Jamaica St., Glasgow.

C.D.C.

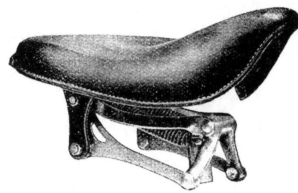

THE ONLY PERFECT SADDLES are XL=ALL'S

Anatomical Cantilever, padded all over, 20/-, 25/- and 30/- each.
Back Rests for same, 7/6, 12/6 and 15/.

THEY are anatomically correct, they spring more evenly and to a greater extent than any other make. They never bounce the rider from his seat and won't wriggle to either side.

COLONEL KENNARD says: "Having had experience of all the best saddles on the market, I am sure your seat is far and away ahead of all others."

MR. A. COX, St. Andrew's Hill, Cambridge, says: "My XL-ALL I would not part with for fifty other saddles. It is the last word in saddles and will soon be all XL-ALL and no other."

JACK GUZZWELL, who broke the **World's 6 Days' Record** by riding 2801 miles—or 467 miles daily—wrote saying: "You have verified your title of XL-ALL. Your pan Motor Cycle Seat is THE BEST. I wish, as you heartily deserve, every success."

Send postcard for 1913 Descriptive List and eight pages of riders' own experiences.

Other Saddles taken in exchange.

FREE TRIALS ALLOWED

to allow riders to test them against any other make for themselves.

XL=ALL LTD.,
Hall Green,
BIRMINGHAM.

Telephone: 38 SHIRLEY.

Bucket Seats, padded, 50/-

The A.B.C. of Dynamo Design.

By ALFRED H. AVERY, A.I.E.E.

An Illustrated Handbook describing General Principles of Dynamo Design. Folding Plate of 500-watt Dynamo. Chapters—I. General Principles of Dynamo Design. II. The Armature. III. Armature Action and Reaction. IV. The Construction and Winding of Armatures. V. The Field Magnet. VI. Design for a 30-watt Dynamo. VII. Design for a 500-watt Dynamo. VIII. Design for a 2-kilowatt Dynamo. IX. Notes on Dynamo Construction and Winding. Price 2s. net ; post free, 2s. 3d.

Small Dynamos and Motors:
How to Make and Use Them.

By F. E. POWELL.

Chapters—I. General Considerations for the Amateur Dynamo or Motor Builder. II. Field-Magnets : Types and Sizes for Various Outputs. III. Armatures : Different Types and How to Wind Them. IV. Commutators and other Details. V. Tables of Windings for Small Dynamos and Motors. VI. How to Build a Small Machine. VII. Useful Data for Dynamo and Motor Builders and Users. VIII. Hints on Testing and Repairing. Price 6d. net ; post free, 7d.

Electric Lighting for Amateurs.

Edited by PERCIVAL MARSHALL, A.I.MECH.E.

Chapters—I. How to Make Small Incandescent Electric Lamps. II. How to Fit up Small Electric Lamps. III. Intermittent Electric Lighting. IV. How to Construct an Electric Night Light. V. A Portable Electric Lamp. VI. How to Work a Small Accumulator Installation. VII. The Control of Electric Lamps from Two or More Points. VIII. An Amateur's Dynamo Electric Lighting Plant. Price 6d. net ; post free, 7d.

Electric Batteries: How to Make and Use Them.

Edited by PERCIVAL MARSHALL, A.I.MECH.E.

Chapters—I. Introductory. II. Types of Batteries. III. How to Make an 8-cell Bichromate Battery. IV. How to Make a Non-Polarising Bichromate Battery—How to Make a Bunsen Cell—How to Make a Daniell Cell—How to Make a Dry Battery. V. Arrangement of Cells in Groups. Price 6d. net ; post free, 7d.

All these books are fully illustrated and give practical advice. Price 6d. each net from all Booksellers, or post free 7d. each from

PERCIVAL MARSHALL & CO.,
66, Farringdon Street, London, E.C.

THE "MODEL ENGINEER" SERIES OF HANDBOOKS.

Price 6d. net each; post free, 7d. each.

No. 1. **Small Accumulators:** How Made and Used. 40 Illus.
No. 2. **The Slide Valve:** Simply Explained. 36 Illustrations.
No. 3. **Electric Bells and Alarms.** 52 Illustrations.
No. 4. **Telephones and Microphones.** Fully Illustrated.
No. 5. **Electric Batteries:** How to Make and Use Them. 34 Illus.
No. 6. **Model Boiler Making.** Fully Illustrated.
No. 7. **Metal Working Tools** and Their Uses. Fully Illustrated.
No. 8. **Simple Electrical Working Models.** Illustrated.
No. 9. **Simple Mechanical Working Models.** Illustrated.
No. 10. **Small Dynamos and Motors.** Illustrated.
No. 11. **Induction Coils for Amateurs:** How to Make and Use Them. Fully Illustrated.
No. 12. **Model Steamer Building:** Hulls and Deck Fittings. Fully Illustrated.
No. 13. **Machinery for Model Steamers.** Fully Illustrated.
No. 14. **Small Electric Motors.** Illustrated.
No. 15. **Simple Scientific Experiments.** 59 Illustrations.
No. 16. **Meteorological Instruments and Weather Forecasts.** Fully Illustrated.
No. 17. **The Locomotive:** Simply Explained. Illustrated.
No. 18. **Simple Experiments in Static Electricity.** Fully Illus.
No. 19. **X-Rays:** Simply Explained. Fully Illustrated with Drawings and Photographs.
No. 20. **Patents:** Simply Explained. Illustrated.
No. 21. **Mechanical Drawing:** Simply Explained. Illustrated.
No. 22. **Electric Lighting for Amateurs.** Fully Illustrated.
No. 23. **Model Steam Turbines:** How to Design and Build Them. Fully Illustrated.
No. 24. **Small Electrical Measuring Instruments:** How to Make and Use Them. Fully Illustrated.
No. 25. **The Beginner's Guide to the Lathe.** 75 Illustrations.
No. 26. **Gas and Oil Engines Simply Explained.** Illus.
No. 27. **A Guide to Standard Screw Threads and Twist Drills.** Small Sizes.
No. 28. **Model Steam Engines.** Illustrated.
No. 29. **Simple Photographic Experiments.** Illustrated.
No. 30. **Simple Chemical Experiments.** Illustrated.
No. 31. **Electrical Apparatus:** Simply Explained. Illustrated.
No. 32. **The Wimshurst Machine:** How to Make and Use It. Illustrated.
No. 33. **Alternating Currents** Simply Explained. Illustrated.
No. 34. **Magnets and Magnetism:** Simply Explained. Fully Illus.
No. 35. **Optical Instruments:** Simply Explained. Illustrated.
No. 36. **Windmills and Wind Motors.** Fully Illustrated.
No. 37. **Wireless Telegraphy:** Simply Explained.

Other Useful Books for this Series are in course of Preparation.

MAY BE ORDERED THROUGH ANY BOOKSELLER.

PERCIVAL MARSHALL & Co.,
66, Farringdon Street, London, E.C.

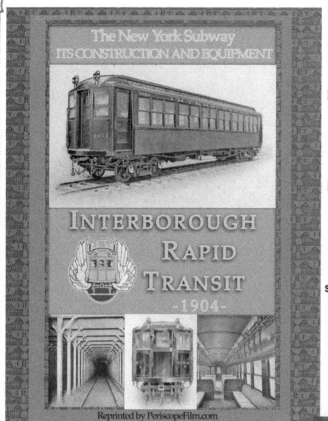

On October 27, 1904, the Interborough Rapid Transit Company opened the first subway in New York City. Running between City Hall and 145th Street at Broadway, the line was greeted with enthusiasm and, in some circles, trepidation. Created under the supervision of Chief Engineer S.L.F. Deyo, the arrival of the IRT foreshadowed the end of the "elevated" transit era on the island of Manhattan. The subway proved such a success that the IRT Co. soon achieved a monopoly on New York public transit. In 1940 the IRT and its rival the BMT were taken over by the City of New York. Today, the IRT subway lines still exist, primarily in Manhattan where they are operated as the "A Division" of the subway. Reprinted here is a special book created by the IRT, recounting the design and construction of the fledgling subway system. Originally created in 1904, it presents the IRT story with a flourish, and with numerous fascinating illustrations and rare photographs.

Originally written in the late 1900's and then periodically revised, A History of the Baldwin Locomotive Works chronicles the origins and growth of one of America's greatest industrial-era corporations. Founded in the early 1830's by Philadelphia jeweler Matthais Baldwin, the company built a huge number of steam locomotives before ceasing production in 1949. These included the 4-4-0 American type, 2-8-2 Mikado and 2-8-0 Consolidation. Hit hard by the loss of the steam engine market, Baldwin soldiered on for a brief while, producing electric and diesel engines. General Electric's dominance of the market proved too much, and Baldwin finally closed its doors in 1956. By that time over 70,500 Baldwin locomotives had been produced. This high quality reprint of the official company history dates from 1920. The book has been slightly reformatted, but care has been taken to preserve the integrity of the text.

NOW AVAILABLE AT
WWW.PERISCOPEFILM.COM